Ectodermal Dysplasias:
A Clinical and Genetic Study

Ectodermal Dysplasias:
A Clinical and Genetic Study

Newton Freire-Maia
Marta Pinheiro

Department of Genetics
Federal University of Paraná
Curitiba, Brasil

Alan R. Liss, Inc., New York

Address all Inquiries to the Publisher
Alan R. Liss, Inc., 150 Fifth Avenue, New York, NY 10011

Copyright © 1984 Alan R. Liss, Inc.

Printed in the United States of America.

Library of Congress Cataloging in Publication Data

Freire-Maia, Newton.
    Ectodermal dysplasias.
    Bibliography: p. 209
    Includes index.
    1. Ectodermal dysplasia. I. Pinheiro, Marta.
II. Title. [DNLM: 1. Ectodermal Dysplasia—familial &
genetic. WR 218 F866e]
QM691.F74 1984   616'.043   84-17183
ISBN 0-8451-0238-9

Q 8.2
F 1

To John M. Opitz,
our friend and teacher:
         a tribute for his charisms.

Contents

List of Figures

The Authors

Newton Freire-Maia, the senior founder of human genetics in Brazil, began life in 1918 in the beautiful old colonial town of Boa Esperança in the interior of the State of Minas Gerais. His paternal grandfather, Domiciano Maia, was a gifted amateur naturalist who had a good laboratory for the study of plants and animals. This is where his grandson was inspired at the age of 15 toward a life-long vocation of research in biology. Before the age of 20 he had already produced a series of essays on Heredity and Life, which according to his friends and admirers revealed his grasp of the subject and a powerful, inquiring mind but which Freire-Maia has purged from his bibliography as being "very horrible," together with several dozens of newspaper articles mostly on science.

Being unable to study genetics at Alfenas, the nearest institution of higher learning, Freire-Maia first became a dentist (1945). Visiting São Paulo, he impressed André Dreyfus, who had founded the genetics laboratory at the University of São Paulo (USP) and who had brought Dobzhansky to the USP to initiate Drosophila studies in Brazil. Freire-Maia dates the beginning of his scientific career from 1946, when he came to work in the USP Genetics Lab on Drosophila genetics. Dozens of discoveries and papers resulted from his Drosophila work over the next few years; however, rather early he was attracted to human genetics, partly inspired by a reading of Dahlberg's *Mathematical Methods for Population Genetics*. His first paper in that field ("Eugenics and population genetics") was published in 1950. Thus, Freire-Maia became the first human geneticist in Brazil. In 1951 he was asked to form the first center for the study of genetics at the Federal University of Paraná in Curitiba. Later this center became the department of genetics in which Freire-Maia worked, also serving briefly as its chairman, attaining emeritus status in 1978. As an outstanding and well-rounded biologist and gifted mathematician, he initiated a series of pioneering consanguinity and population genetics studies. He defended his thesis for the degree of doctor in natural sciences at the Federal University of Rio de Janeiro (then still the University of Brazil) with his data on the effect of inbreeding on the Brazilian population.

Freire-Maia's distinguished career at Paraná was interrupted by a year (1956–1957) of studies with Neel at Ann Arbor and by a year (1970–1971) as Scientist in the Genetic Unit of the WHO in Geneva, where he began his systematic studies of the ectodermal dysplasias. In Curitiba, Freire-Maia formed a school of human genetics that produced such outstanding scientists as Henrique Krieger, his brother, Ademar Freire-Maia, his wife, Eleidi A. Chautard-Freire-Maia (also a student of John Edwards), Francisco A. Marçallo, A. Quelce-Salgado, and many others who now work in several different states of Brazil. Freire-Maia has represented Latin America on the Organizing Committee of the International Congresses of Human Genetics; he is a founder and was president of the Brazilian Society of Genetics, and a founder and First Vice-President of the Brazilian Society for the Progress of Science. A world traveler, outstanding teacher and accomplished linguist, Freire-Maia is internationally known and honored for his work in Drosophila population, radiation, and anthropological genetics, consanguinity, limb anomalies, and the ectodermal dysplasias. Several hundred publications include textbooks entitled *Medical Genetics* and *Genetic Effects of Radiation in Man* (both with A. Freire-Maia), *Problems in Human Biology—A Study of Brazilian Populations* (with Francisco M. Salzono), *Human Radiation Genetics*, *Human Population Genetics*, *Topics in Human Genetics*, and six editions of the book *Brasil—Laboratório Racial*. It was, therefore, a foregone conclusion that I would ask Freire-Maia to be a Brazilian member of the distinguished Board of Corresponding Editors of the American Journal of Medical Genetics. Freire-Maia is a full member of the Brazilian Academy of Science.

The section on genetic counseling in this book shows forth Freire-Maia's deep humanity. He is one of the kindest and most caring of persons known to me and a wonderfully affectionate friend to many people. The wellsprings of his ethical-moral views in personal and professional life are Franciscan rather than Marxist-Socialist. Freire-Maia is a scholar and humanist of great culture, deep feeling, and with a marvelous sense of humor. I am additionally grateful to him for making possible my first visit to Brazil, a country that has made giant contributions to human and medical genetics, inspired largely by leaders such as Freire-Maia.

Marta Pinheiro, the co-author of this book, is a brilliant, young, already accomplished geneticist who obtained her undergraduate diploma in biology and her master of science degree in human genetics at the Federal University of Paraná (under N. Freire-Maia). Her master of science thesis, a model of scientific precision and elegance, dealt with clinical and genetic studies of two large Brazilian kindreds with several members with Christ-Siemens-Touraine syndrome, including an analysis of the manifestation rate among carriers. Marta Pinheiro obtained her PhD degree in human genetics at the University of São Paulo (under O. Frota-Pessoa) with a thesis on the ectodermal dysplasias. She is a researcher of the Brazilian National Council for Scientific and Technological Development and works full time at the Unit of Medical and Population Genetics (headed by N. Freire-Maia) of the Department of Genetics of the Federal University of Paraná, where she works with the senior author.

John M. Opitz

Foreword

In this book the authors do far more than discuss the ectodermal dysplasias. They also offer an overview of the nomenclature of congenital anomalies and take another turn at grappling with the classification of "diseases," which has been a major epistemological problem of western medicine since the efforts of Sydenham, Isidore Geoffroy St. Hilaire, and many others, especially since the introduction of modern taxonomic methods into zoology and botany by Linnaeus. By now seasoned experts at the ectodermal dysplasias, the authors, Newton Freire-Maia and his collaborator, Marta Pinheiro, also take a retrospective look at the historical development of the concept "ectodermal dysplasia." They then discuss each of the 108 clinical entities they have chosen to include (a total of 117 due to heterogeneity) and which they have organized in a nosologically most useful manner. I like the uniformity of the style of the entries—at a glance it permits an appraisal of what is known, but even more importantly, of what is *not* known about the entity, something usually not possible to discern in treatises that discuss the conditions in prose form. Having presented each disorder separately, they then sum up in the best clinical genetical manner by emphasizing causal heterogeneity.

For those seeking a key to genetic usage, I would recommend reading first the two last chapters, in which the authors explain with meticulous care the evidence on which they base their genetic inferences. These chapters are no less commendable for the humane and caring view they give of genetic counseling, a view that evolved in Freire-Maia over a 30-year span of being the sole human geneticist and genetic counselor in a large region of Brazil, a country that at this writing still severely limits the reproductive options available to its citizens.

Throughout the book one is properly reminded of the fact that skin is an organ (not just a tissue) and that ectoderm gives rise to a whole series of organs including skin, central nervous system (CNS), sense organs, and neural crest derivatives. And while most of the entities discussed in this book undoubtedly contain a component of or are entirely dysplasias, ie, tissue disturbances (defects of histogenesis), many can be considered syndromes since they also contain malformations, ie, defects of organogenesis. Besides the axial and paraxial metameric organization that gave it stability and a basic structural unity and developmental plan, the second most important attribute of our vertebrate lineage was the development of the outer covering of the body, which initially kept salt out, and later, after the beginning of tetrapodial terrestrial life, kept water in.

During various phases of vertebrate evolution, ectoderm, besides the basic outer covering, gave rise to a dazzling variety of organs for defense, including the horns of the rhinoceros and goats and their relatives (including the North American pronghorn), tail spikes of dinosaurs, claws of cats, spines of echidna and hedgehogs, quills in porcupine, girdle scales in armadillos, poison spur of platypus, and poison glands in frogs and toads. Other ectodermal derivatives in various vertebrate lines include mechanisms for temperature regulation; special skin patches to incubate eggs of birds; pouches to protect the immature newborn of marsupials; the supremely important organs of lactation, which Linnaeus recognized as the *sine qua non* of mammals; climbing and swimming aids such as claws or the wing hook of the Hoatzin, and webbed toes of many mammals, birds, and amphibians; proprioceptive and other sensory organs, such as the lateral line organ of fish; horny plates in reptiles; wet, mucous skin in amphibians; feathers in birds; sebaceous and sweat glands in humans and various oil glands to protect skin and feathers; sexual attractants such as wattles of birds and vivid coloration of anogluteal skin of monkeys; and finally the most important guarantor of survival during mammalian evolution—the presence of fur. Here is but a very limited enumeration of vertebrate ectodermal/epidermal specialization, but perhaps it is sufficient to hint at the supreme importance of this organ system to survival, quality of life, and reproduction.

Phylogenetically, humans are sadly regressed with respect to epidermal derivatives; claws have become nails and an aggravation to every child having to clean them; most of the fur is gone; several pairs of mammary glands have been reduced to the sole pair of the human female; and human bottoms and faces can never hope to match those of a mandrill or hamadryas baboon in full glory. While important, though not a necessity, under conditions of modern life, sebaceous glands have become an emotional burden and health hazard to millions of adolescents with acne, just as teeth are rotting in persons formerly unacquainted with toothbrushes but also never exposed to refined carbohydrates. If it is any consolation, it must be recorded that baldness is not just a human liability but, to a variable extent, an attribute of all primates.

Ontogenetically, though, it is important to come back to a vital point, namely the fact of skin as *organ* development. By definition, organogenesis means field development, fields being the embryonic regions in which the development of the complex structure appropriate to that region is controlled and regulated in a temporally synchronous, spatially coordinated and epimorphically hierarchical manner, ie, in a genetic plan generic for a species but varying slightly from individual to individual depending on race, familial variants and environmental factors. Since anatomical and histological structure is mute about the events that shaped it, and since, with rare exceptions, an experimental embryology of humans is impossible, indirect evidence for fields must be sought.

In humans, the most persuasive evidence is the occurrence of the *same* malformation in individuals with conditions of *different* cause. This is because identical structure means identical development regardless of initial causal mechanism. Human organs are evidently capable of responding to a huge number of diverse dysmorphogenetic causes with the production of only a very limited repertoire of malformations—precisely because coordinated field development limits independent response from component structures. From this it also follows that there is no such thing as a causally specific malformation. And while the entities in this book are generically referred to as "dysplasias," many of them, as the authors rightly point out, are in fact malformation/dysplasia syndromes, with less of a risk of neoplasia than is seen in many other dysplasias and dysplasia syndromes. To the thoughtful reader a careful perusal of the entities in this book will uncover many a previously unrecognized developmental field relationship.

Mutually inductive relationships between ectoderm and mesoderm must be remembered; if a limb defect (eg, ectrodactyly) is present in an ectodermal dysplasia syndrome, the apical ectodermal ridge was involved. Since this special type of ectoderm surrounds the embryo as a ring (Blechschmidt's *Ektodermring*), which extends from oropharyngeal areas over limb buds to tail (Grüneberg's tail ridge), one would not be surprised to see in such a syndrome variable oral and dental involvement, and occasionally perhaps an imperforate anus. Most of the conditions discussed in this book require extensive further studies to delineate their complete phenotypic spectrum, to obtain a better insight into pathogenesis, and to define cause and possible heterogeneity. The format of this book makes these needs easily identified.

This book could not have been written by a better team: As a student of Dr. Freire-Maia, Dr. Pinheiro is steeped in her teacher's methods and has collaborated in many of his studies, discoveries, and publications. The senior author wrote his first paper on a genetic skin disorder in 1967 and on an ectodermal dysplasia syndrome in 1970. During his stay at the World Health Organization (WHO) in Geneva in 1970–1971, he began the systematic studies of the ectodermal dysplasias that have resulted in this book and have made him the world authority on ectodermal dysplasias. Freire-Maia has described more new ectodermal dysplasias than anyone else.

This book is a unique pioneering effort and impressively fills a great void. I commend it to all clinical geneticists, dermatologists, and developmental geneticists and pathologists with an interest in skin and ectoderm. I vigorously endorse the authors' hope about Curitiba's becoming a world reference center on the ectodermal dysplasias; indeed I would make a strong recommendation that all pertinent new or diagnostically doubtful material be referred to Drs. Freire-Maia and Pinheiro for diagnosis (if possible), and inclusion in their registry to speed identification of new disorders.

This book arrives at a propitious time not only from a genetic-biological but also from a human point of view. I received a letter from Mrs. Mary Kay Richter from Trenton, Illinois, recounting the poignant story of her son Charles who evidently has the Christ-Siemens-Touraine syndrome, and of her strenuous efforts to get a diagnosis and some help to aid temperature regulation and to fit the lad with some dentures to assist in developing normal speech and eating habits. The frustration that she, like many other intelligent, energetic mothers in similar circumstances experienced, prompted her to found the National Foundation for Ectodermal Dysplasias*, (the plural form of dysplasia indicating an astute recognition of heterogeneity). This book will surely provide plentiful information on that subject, and in the hands of competent, well-intentioned professionals it will obviate much of the diagnostic frustration experienced by this mother. May all other professionals then rise to provide proper prognostic, genetic, and habilitative information and the overall medical and psychological care that ectodermal dysplasia persons need to such an exceptional degree.

John M. Opitz
Basin, Montana

*108 North First, Suite 311, Mascoutah, IL 62258.

Preface

Ectodermal dysplasias are a large and heterogeneous group of conditions. Each year new entities are described and labelled as ectodermal dysplasias. Several years back, in an attempt to reduce the nosologic chaos in the field, Freire-Maia [1971] proposed a clinical definition and a clinical-mnemonic classification of ectodermal dysplasias. Since that time, several other reviews have been published [Freire-Maia, 1973, 1977a, 1982; Settineri, 1974; Witkop et al, 1975; Rapone-Gaidzinski, 1978; Freire-Maia and Pinheiro, 1983b, c], a "community of diseases" has been described within ectodermal dysplasias [Pinsky, 1975], and the group was enlarged to include many "new" conditions.

This book is a review of ectodermal dysplasias *sensu lato,* ie, both pure ectodermal dysplasias and ectodermal dysplasia/malformation syndromes. It widens the scope of some of our previous papers and, obviously, is partly based on them. For example, Chapters 1 and 2 are based on Freire-Maia [1977a, b], but contain a number of additions and adaptations.

Each condition will be described only once, even when heterogeneity (on etiological grounds) is suspected. The most commonly used or clinically best designation will be adopted in each case. Obviously, our preferences may differ from those of some of the readers. Thus we regret if the nomenclature used does not satisfy all readers. This would be simply impossible. Therefore, eponyms and descriptive terms (some of them with their respective acronyms) will be used alternately throughout the book. In each case we tried to find terms that seemed to be most appropriate to us. When an appropriate term was not found, we preferred to use the easiest alternative—the eponym.

Practically all types of designations proposed thus far have been criticized. The nomenclature includes etiology, eponyms, acronyms, names of cities, initials of first patients, and clinical descriptions (either alone or followed by numbers, geographical designations, names of authors, etc.). An analysis of the nomenclature is outside our scope. The reader may find detailed analyses of this topic in Opitz et al [1969], Warkany [1971], Herrmann and Opitz [1974], Freire-Maia [1976], Herrmann et al [1977], Herrmann [1979], and Spranger et al [1982]. Eponyms and clinically decriptive terms are the most frequent forms of nomenclature used in this book. This reflects what actually happens in the field of syndromology, except for the so-called chromosome disorders that are not present among ectodermal dysplasias.*

The corresponding number in the fifth edition of McKusick [1983] will be presented after the name of each condition. When the condition is not listed in McKusick's book, it will be denoted "McK: not listed."

We are very grateful to the National Council for Scientific and Technological Development (Brazil) and to the World Health Organization (Geneva) for a number of grants that supported our investigations. The senior author is particularly indebted to WHO, in whose headquarters in Geneva he had the opportunity to develop his concept and classification of ectodermal dysplasias in 1970-1971 during his stay at the Human Genetics Unit, directed at that time by his good friend, Dr. Italo Barrai.

We also acknowledge the assistance of Dr. John M. Optiz, who read the manuscript and made a number of valuable scientific corrections and suggestions to improve the text. He was also kind enough to honor this book with a preface. However, since English is not the mother-tongue of the authors, the work of Dr. Optiz encompassed more than the usual scientific analysis of the text, an area in which his help was of the utmost importance. Our text was substantially improved both in terms of scientific level and English style, with the aid of his editorial pencil. However, since Dr. Opitz could not possibly check all the information contained in this book, any mistakes still present are strictly our responsibility.

Our thanks are also due to Mrs. Raquel Rapone-Gaidzinski for suggestions at the beginning of our work, to Drs. Eleidi A. Chautard-Freire-Maia and David Carneiro, Jr., for reading and criticizing Chapter 15, to Miss Irene Sedoski for typing the manuscript, to Mrs. Elettra Greene for a final review of the text, and to all of the authors who kindly sent us photographs of patients as well as to a number of publishers and journal editors for permission to reproduce these photos. Credits to the authors appear in the figure legends and credits to the journals in the "List of Figures." Unpublished photos will be referred to as such, even when the patients have been described before.

In 1981, we created in our department a center for the study of ectodermal dysplasias that functions, since that year, as an international reference center for this group of conditions. We have already received consultations from colleagues of 12 countries of America, Europe, and Africa and are willing to widen the geographic distribution of our correspondents, in the hope that we might become helpful to a larger number of persons from a larger number of countries. We are interested in receiving clinical data and photos of patients with ectodermal dysplasia and to study them in order to

---

*Which is not to say that ectodermal dysplasias cannot occur in aneuploidy syndromes; on the contrary, all aneuploidy syndromes are malformation/dysplasia syndromes.

help all colleagues interested in the field to obtain better knowledge and to provide their patients with better care. This is one of the main purposes of our center.

Aware of the fact that this book is open to criticism, the authors would be very grateful to any reader who could find time to send them corrections and suggestions.

Newton Freire-Maia
Marta Pinheiro
Dept. of Genetics
Federal University of
  Paraná
Caixa Postal 19071
8000 Curitiba, PR, Brasil

1. Introduction: Nosologic Groups

A given disease, dysplasia, anomaly, or syndrome may be clinically and causally delineated from others that are similar (even when some phenotypic overlap exists among them) if enough patients are available for study and sensitive techniques of phenotypic and genetic analysis are used.

However, the organization of different conditions into groups is often based on criteria of convenience. Thus, a given condition may be placed in one or another group solely according to the signs or symptoms arbitrarily chosen as the criteria for membership in that particular group. Such groups may contain rather different entities whose phenotypic overlap is superficial. However, in some instances, presumed pathogenetic similarity is the essential basis for the creation of a nosologic group, which is likely to contain rather similar entities with multiple and intimate phenotypic overlap.

A nosologic group contains two or more conditions sharing one or more clinical manifestations. Different conditions may be classified as a group for a number of reasons. They may be characterized by skin bullae, or they may share defects both in the skin and in the nervous system, or the patients may have short stature, or exhibit signs in structures deriving from the ectoderm, etc. Thus, groups of conditions such as keratoses, epidermolyses, clotting disorders, phacomatoses, dwarfisms, hereditary mucosal syndromes, chromosome breakage syndromes, and ectodermal dysplasias are broad groupings that generally may be enlarged or reduced according to the convenience or purpose of the author.

Let us consider the so-called chromosome breakage syndromes, a group that includes Bloom's syndrome, Fanconi's anemia, Louis-Bar syndrome, xeroderma pigmentosum, and a few other conditions [German, 1969, 1972]. They are characterized by growth retardation, tendency to malignancy, and to an abnormally high frequency of chromosome rearrangements. Since the biochemical processes related to these peculiarities are not known, German [1969] admits that "the lumping of these conditions may thus be advantageous for certain purposes (eg, those of the cellular

1

molecular biologists or the oncologist) while disadvantageous for others (eg, the clinician concerned with metabolic and therapeutic problems)" (page 129). This reflects the fact that, as mentioned, classifications are sometimes based on convenience, which may differ from one group of specialists to another.

Since the unit of study is not the group but the condition itself, the same condition may be classified under different headings, depending on the interest of the investigator. The Ellis-van Creveld's syndrome (a type of chondrodystrophic dysplasia), oculodentodigital dysplasia (a craniotubular bone dysplasia), and epidermolysis bullosa, for example, have all been classified as ectodermal dysplasias by Freire-Maia [1971] and Witkop et al [1975]. However, the condition itself is already a major unit, since the real basic working unit is the patient: Patients with the same condition do not generally exhibit identical manifestations, and the apparently identical conditions in unrelated patients may be due to two or more different causes.

Attempts to classify different conditions into nosologic groups on the basis of clinical, biochemical, causal, and other similarities for teaching, diagnostic, investigative purposes, etc, should not be confused with more pretentious objectives such as that of McKenzie [1958], who created the designation "first arch syndrome" to encompass a group of "anomalies" affecting the head and the neck that arise from "abnormal development of the first visceral arch," ie, Treacher Collins' syndrome, Pierre Robin's syndrome, Waardenburg's syndrome*, cleft lip, cleft palate, congenital deaf mutism, hypertelorism, etc. McKenzie claims that "all the anomalies mentioned comprise one hereditary syndrome [the first arch syndrome] caused by a dominant gene or group of dominant genes with variable specificity and expressivity and with only a moderate degree of penetrance" (page 485). Penetrance would depend on "the nutritional state and diet of the mother during the first weeks of pregnancy" (page 484). The expressivity and specificity of the gene(s) would depend "on the details and timing of the maladjustments occurring among the vessels concerned" (page 484). According to this concept, therefore, the "first arch syndrome" not only includes a group of malformations and "syndromes" affecting a

---

*Note that the so-called Waardenburg's syndrome has been found to be a complex of two different conditions (Waardenburg's syndrome I and II; cf, among others, Arias [1971, 1980]; for other references, see McKusick [1983] entry no. 19351). According to several authors, the syndrome described by Klein [1950] is a different condition; it should be called Klein's syndrome.

specific region (head and neck), but—much more than that—intends to be a detailed (*and totally unproved*) genetic and pathogenetic theory relating all of them to the same basic factors [see also McKenzie, 1968]. McKenzie's concept of first arch syndrome should be totally abandoned as nonsensical.

Reflecting on a widespread tendency among teratologists, clinicians, geneticists, etc, to lump entities together, Warkany [1971] affirms that it is understandable that the teratologic taxonomist has a desire to classify syndromes in order to reduce the "apparent chaos" that emerged as a result of the accelerated recognition of new malformation syndromes. Pinsky [1974, 1975] suggests that one way to reduce the taxonomic chaos in which most dysmorphic syndromes are involved is to incorporate as many of them as possible into "phenotypic communities" of clinically similar conditions regardless of their cause. This means that syndromes due to oligogenes, to chromosome aberrations, to exogenous factors, etc, may be equally entitled to membership in the same community. This is different from the abusive incorporation of different nosologic entities into a single syndrome (see, for example, the concept of first arch syndrome) and from dubious concepts reflected in the use of expressions such as "forms of transition," "incomplete syndromes," "syndrome A is a mild form (or variant) of syndrome B," etc. McKusick [1979], in speaking against the use of such vague and misleading expressions, correctly says that "phenotypic overlap (of diseases) is not necessarily grounds for considering them fundamentally the same or even closely related" (page xi). Pinsky's position is that syndromes with multiple foci of phenotypic overlap should be recognized as forming "communities." These communities may then be partitioned into "subcommunities" as he did, for example, for the hand-foot-ectodermal dysplasias [Pinsky, 1975]. Some ectodermal dysplasias include distal limb malformations while others do not. The community with limb anomalies was subdivided into two groups: those with anomalies of the alae nasi or labial region and those without these malformations. The possible dysmorphogenetic significance of this classification is supported by the fact that sensorineural deafness is uncommon in the first subgroup while it is relatively common in the second. Pinsky's approach, therefore, is more than a simple grouping of conditions that appear vaguely similar, either because they all share some specific anomaly ("leitmotiv") or simply because they share malformations in the same organs. Communities of syndromes are more than mere aggregates of conditions that share clinical similarities or affect the same body region since phenotypic similarities within "communities" *may* reflect pathogenetic (ie, dysmorphogenetic) similarities. By analyzing their patterns of clinical similarities and dissimilarities, it may be possible to generate working hypotheses about the general pathogenetic basis or bases for their similarity. However, it is important to remember that, until such hypotheses are tested and corroborated, a "community" represents nothing more than a nosologic device that may be useful in areas such as differential diagnosis, teaching, research, and information retrieval. Thus, extensive clinical resemblance *may* reflect pathogenetic resemblance, but the former is never, per se, a proof of the latter.

Some well-known nosologic groups, such as the mucopolysaccharidoses, are good examples of communities that are not artificial and whose creation was not regulated by pure convenience. Mucopolysaccharidoses are a real family of diseases because of the pathogenetic similarities among them. Other groupings, such as "limb defects," "limb deficiencies," "congenital heart diseases," "cancer," "clotting disorders," are more artificial. Thus, nosologic groups may have different degrees of artificiality: mucopolysaccharidoses are located at the low end of the range, while others, such as "limb defects," are at the high end. Ectodermal dysplasias in general can be placed somewhere between the two extremes, though some subgroups of ectodermal dysplasias would probably occupy a position closer to mucopolysaccharidoses than to that of ectodermal dysplasias as a whole.

To sum up, in a world where everything is classified and subdivided into groups (from stars and nations to people and home gadgets), classifying seems to be a natural tendency of the mind. Nosologic groups, even when they are more artificial than they should be fulfill a need from the viewpoint of research, teaching, differential diagnosis, and bibliographic retrieval. This is an operational concept. In this sense, nosological groups may always prove to be useful. Even those merely based on the location and type of developmental error (deficiency defects of the limbs, amauroses, congenital heart diseases, etc) are important in the sense that they designate fields of specialization and therefore circumscribe areas of study and work. However, nosologic grouping must progress from mere accumulation of vaguely and superficially similar conditions to communities (or families) of conditions, whose components share profound and multiple clinical overlaps that may be suggestive of pathogenetic similarities.

2. Definitions and Classifications

DEVELOPMENTAL EMBRYONIC FIELDS

The concept of the developmental field stems fundamentally from the work of the German embryologists Boveri and Spemann and was introduced into medical genetics by Opitz [cf Opitz, 1979, 1981a,b; Opitz and Gilbert, 1981; Spranger et al, 1982].

The first blastomeres are equipotent and totipotent, ie, they are not locally differentiated; therefore, any one of them has the potential of taking any developmental direction. The fertilized egg and these first blastomeres are called the "primary field." During the human equivalent of gastrulation, cells in different areas of the embryo undergo specific and irreversible differentiation. These areas are called secondary or epimorphic developmental fields, and their growth processes are controlled and coordinated in a spatially ordered, temporally synchronized, and epimorphically hierarchical manner. By "epimorphically hierarchical manner," we mean a sequential developmental progression from a given stage to a more highly complex and mature stage. Epimorphic development gradually limits and reduces the size of the field on which any intrinsic or extrinsic dysmorphogenetic cause can act. Borders of different fields may overlap and interact. Components of a field may be closely contiguous or more distantly located. The field is said to be monotopic in the first instance and polytopic in the second.

Evidence for the existence of developmental fields is found in clinical genetics, teratology, comparative anatomy, experimental embryology, etc [Opitz, 1981a,b; Opitz and Gilbert, 1981].

Some practical consequences may be inferred from the field concept:

1) A single cause acting on a single field produces, by definition, a single but complex malformation involving all the structures normally derived from that field.

2) Since some inductive relationships act over a distance (eg, from mesonephros to limb, from gonad to external genitalia), components of a polytopic field defect may be located at a distance from each other.

3) In different individuals, the same cause may act at different times on a given field and produce anomalies differing in type and extent (eg, autosomal dominant holoprosencephaly producing cyclopia when acting early, microcephaly with borderline intelligence and unilateral cleft palate when acting later, and normal head size and intelligence but single upper central incisor at an even later stage).

4) Different extrinsic or intrinsic causes acting on the same field may produce the same malformations. ("There is no such thing as a causally specific malformation" [Opitz, 1981a; page 12]).

5) A dysmorphogenetic *process* in operation in a given field may affect other contiguous or noncontiguous fields, thus producing complex conditions representing *relational* pleiotropy.

6) A given dysmorphogenetic *cause* may affect more than one field, thus producing syndromes representing *mosaic* pleiotropy.

These six consequences of the field concept are basic in medical genetics.

The distinction between monotopic and polytopic fields is not so simple as it may appear from a superficial analysis of the problem. Hanhart's "syndrome," although characterized by such distant signs as micrognathia and deficiency of the distal part of at least one limb, can be considered a monotopic field defect when viewed from an embryological perspective. Micrognathia and limb defects derive from disturbances of the ectodermal-mesodermal interactions in the area of the ectoderm ring that involves part of the face and limbs. However, the acrorenal anomalies are a group of polytopic field defects because they do not derive from a single contiguous embryonic primordium but from two separate ones that, in spite of the distance separating them, keep an intimate developmental relationship. These examples raise the problem of the long-distance effect in embryological development. At the beginning of embryogenesis, mutually inductive influences occur in local cell groups, but, as embryogenesis continues, long-distance influences are expected to occur as they really do. Returning to one of the above examples, limb buds and the primordial renal system are rather close during early embryogenesis, and the fact that some time later limbs and kidneys are distantly located does not alter the intimate developmental relationship existing between them. (For further information and bibliography, see Opitz [1981], and Opitz and Gilbert [1981].)

It may also occur, as we shall see later, that a single anomaly triggers a cascade of other anomalies. A complex of pathogenetically related anomalies has been called "single syndromic anomaly" by Smith [1969], "anomalad" by a group of specialists [Smith, 1975], and "paradrome" by Freire-Maia [1977b]. However, none of these authors had a clear and precise understanding of the multiple processes involved. This problem will be analyzed later.

The term "anomalad" was coined by Dr. F. Clarke Fraser through the junction of *anomaly* and the suffix *ad,* (as in diad, triad, etc) [Smith, 1975]. The term was proposed during a "frustrating meeting" [Opitz, 1981a] and has been heavily criticized. After a short, precarious life, the term seems to have been definitely deleted from the scientific glossary (for the bibliography on the problem, see Bernirschke et al [1979]. We hope that, excluding historical works, the present book will be the last one in which this unfortunate word appears.

After that "frustrating meeting," a less frustrating one was held. Its recommendations contain contradictions, redundancies, imperfect definitions (based on exclusions), and suggestions against well-established meanings. They are summarized by Spranger et al [1982]. We shall not follow all the recommendations suggested by this international working group but shall propose a compromise between them and what we think is more plausible and correct.

MALFORMATIONS AND DISRUPTIONS

Malformations may be primary or secondary. A *primary* malformation is a morphologic defect of an organ, part of an organ or a larger region, resulting from the action of a single cause on a single developmental field; it represents an *intrinsic* defect of the anlage (primordium). The anomaly may be ascertained at birth (eg, cleft lip, acheiria, polydactyly) or at a later time (eg, pyloric stenosis, small tracheo-esophageal fistula, a congenital heart defect, unilateral renal agenesis, agenesis of corpus callosum). This definition excludes metabolic disturbances (eg, phenylketonuria), dysplasias, and genetic diseases of late onset (abiotrophies; eg, Huntington's chorea).

The expression "congenital malformation" is used so widely that any attempt to change it would be doomed to failure, at least at the present time. By definition, malformations are congenital. If they are not diagnosed at birth, this is assumed to be due to secondary reasons (either because of the physician or the malformation itself).

Secondary malformations (disruptions) are morphological defects resulting from an *extrinsic* action on originally normal developmental fields.

Disruptions may be produced by radiation, drugs, maternal-fetal metabolic disorders, infections, immunological mechanisms, oligohydramnios, amniotic bands, myoma uteri and uterine malformations, ischemia, twinning, hyperthermia, etc. (For an ample analysis of all these factors, see Opitz and Gilbert [1981].) Thus, X-rays, thalidomide, alcohol, diabetes, maternal phenylketonuria, rubella, toxoplasmosis, oligohydramnios, amniotic bands, twinning, hyperthermia (from infections to sauna baths), etc, are disruptive agents. *

Malformations are generally named or classified according to the part involved and the type of defect. Examples are patent ductus arteriosus, anencephaly, bilateral acheiria, left cleft lip, bilateral anotia, aplasia or hypoplasia of the upper lateral incisors. They are mostly anomalies of incomplete or no differentiation (eg, ventricular septal defect, bilateral renal agenesis); less commonly, they are due to abnormal differentiation (eg, extra thumbs). They are also classified as "severe" (those that are incompatible with life or seriously affect the patient's activities) or "mild." The following are included in the severe group: anencephaly, acheiropodia, tetraphocomelia, spina bifida, hydrocephaly, cleft lip and palate. The less and least severe forms of primary malformations ("mild malformations") are still malformations, ie, they are all due to one or more defects of morphogenesis, and, although mild, are always abnormal. Absence of the palmaris longus muscle or of the upper lateral incisors, though apparently selectively neutral, clinically benign traits, are still malformations and not minor anomalies.

MINOR ANOMALIES

The term "minor anomalies" is the antonym of "normal developmental variants," which are all of those quantitative morphologic characteristics that constitute the physical individuality of a person but at the same time are the heritage of the family and ethnic group. In *normal* individuals these variants represent the result of variable interaction between polygenes and

---

*Malformations should not be confused with *deformations,* which are the result of abnormal morphogenesis due to the action of unusual mechanical forces. Deformations may be intrinsic (when the causative forces are due to intrinsic problems of the embryo or fetus, such as a malformation) or extrinsic, when produced by mechanical forces that are extrinsic to an otherwise normal concept. A deformation may be either a single localized process (eg, clubfoot) or a sequence of processes (eg, breech presentation sequence). For an excellent review of deformations, see Smith [1981].

environment and can be considered morphologic "fine-tuning" occurring at the end of, and *not* during, morphogenesis. Disturbances of this process produce metric abnormalities, ie, traits whose measurements are $\leq$ 2–3 standard deviations (SD) beyond the mean of the population distribution. Though abnormal (in this statistical sense), these traits can only be called *minor anomalies,* not mild malformations—since they are *not* defects of organogenesis. Defects of organogenesis, however mild, are borderline defects, ie, traits that, even in their mildest form, do not shade into normality (on these problems, see Opitz [1981b]).

By far the most common abnormalities in aneuploidy syndromes are minor anomalies. Morphologically they are indistinguishable from the normal variants that occur in the nonaneuploid population, hence they must be developmentally identical. All normal variants may occur as minor anomalies, and vice versa. The difference between the aneuploid and nonaneuploid states is that, in the former, extreme variants occur at higher frequency and in specific combinations in single individuals who largely lack galtonian family resemblance, ie, whose first-degree relatives more closely resemble the nonaneuploid population mean.

Among the commonly observed minor anomalies are some degree of ptosis (most commonly of the left upper eyelid), epicanthal folds, a short finger, broad hand, large protruding ears, four-finger crease of palm, ocular or intermammillary hypertelorism, mild hypertrichosis, Darwinian tubercle, wide gap between toes I and II, discrete facial variations, and dermatoglyphic variants.

A number of minor anomalies represent single findings among normal persons and do not seem to be associated with any pathological consequence. These small "defects" are generally assumed to be normal variants. When they are found in patients with multiple malformations, they may be described as members of the large constellation of defects although they may only represent coincidental findings without any causal and/or pathogenetic relationship.

Of course, not only structural but also functional, maturational, and growth characteristics can vary normally. Before assuming, for example, that retarded growth is part of a syndrome, it is necessary to determine whether or not it is a normal trait in the family. For instance, the pure ectodermal dysplasia we described under the name of alopecia-onychodysplasia-hypohidrosis-deafness [Freire-Maia et al, 1977] was observed in a girl whose nonconsanguineous parents were very short. The same applies to hypohidrosis. It is well known that sweating capacity varies largely from person to person. With the technique we used [Cat et al, 1972], normal children produced an average of 208 $\pm$ 11 mg with an SD of 71 mg. However, when we find patients with signs of ectodermal dysplasia who sweat less than this average minus twice its SD or much less than the single values determined among normal relatives, we are inclined to classify the condition as hypohidrotic. We had done so in the case of what has been called odontotrichomelic hypohidrotic dysplasia [Cat et al, 1972]. However, further investigation on the two patients with this syndrome showed that their sweating capacity could be assumed to be normal [Rapone-Gaidzinski, 1978]. Their values were 40 mg and 61 mg in the first determination, against 143 mg and 194 mg for their normal sibs. However, Alcântara-Silka [1977], in an investigation of 304 normal Caucasian persons of both sexes and different age groups (from infancy to old age), detected averages of 160.2 (SD = 59.4) for males 12 to 20 years old, 158.2 (SD = 57.2) for females 3 to 11 years old, 157.0 (SD = 60.5) for males 21 to 33 years old, etc. Extreme averages of 120.6 (SD = 46.9) and 179.4 (SD = 41.9) were found in the 12–20-year-old female group and the 36–59-year-old male group. Individual measurements led to a number of results below 50 (such as 23, 24, 27) and above 200 (such as 253, 251, 249, 245). On the basis of these data, it seemed more reasonable to assume that our two patients were not pathologically hypohidrotic; the name of the condition was then changed to odontotrichomelic syndrome (since it is really a dysplasia/malformation syndrome).

Our experience in this field also involved a case of apparent intermammillary hypertelorism. In a 7-year-old white girl with alopecia-onychodysplasia-hypohidrosis-deafness [Freire-Maia et al, 1977], our first impression was that the patient had an increased intermammillary distance. This was not mentioned in the paper since the hypertelorism did not seem large enough to have any clinical significance. Later on, we compared the measurements of the patient (thorax circumference at the nipple level, 55.0 cm; intermammillary distance, 14.5 cm; ratio, 0.264) with the averages of 50 suitable controls (59.2 cm and 14.1 cm, respectively, ratio 0.238, SD = 0.045). Therefore we concluded that the first impression was an illusion destroyed by quantitation of the trait [Freire-Maia et al, 1981].

Far be it from us to minimize the importance of minor anomalies. Sometimes they really are signs of a syndrome and, in such a situation, only investigation of other patients and unaffected relatives can prove it. In other cases, evident minor anomalies may be associated with major ones not so easily observed [Opitz et al, 1969]. In such cases, minor anomalies are always to be accepted as an invitation to more thorough clinical and laboratory investigations.

Sometimes, a peculiar facial appearance is seen in normal persons. However, any child born with a peculiar appearance should require particularly careful scrutiny. His or her parents, sibs, grandparents, etc, should also be examined, since the trait may be familial.

All of these considerations point to the possibility that many syndromes and dysplasias are being described incorrectly. If the authors lose interest in them and if the conditions are sufficiently rare so as not to be found again soon after discovery, the "new" condition may remain associated with an incomplete or distorted clinical picture for a long time (or even forever, if the syndrome appears on the stage just once and then disappears behind the wings).

MULTIPLE MALFORMATIONS

As mentioned earlier, a single malformation is the result of the action of a single cause on a monotopic developmental field (eg, cleft lip, acheiria, hypospadias). Multiple malformations (or defects) may represent the following:

1) Action of a single cause on a polytopic developmental field leading to pathogenetically correlated defects. The acrorenal malformations are a polytopic field defect.

2) Action of a single cause on a single developmental field leading to a single defect that triggers a cascade of secondary defects in subsequent morphogenesis (sequence). This entire constellation of defects stems from only a single underlying defect that may be a malformation or a dysplasia. The pattern of complication triggered by myelomeningocele (lower limb paralysis, urinary tract infection, renal damage, constipation, and dilatation of the bowel, etc) should be called the myelomeningocele sequence. Aplasias and hypoplasias of the radial ray may lead to a secondary malformation of the contiguous ulna that appears to be short, thick and bent [Freire-Maia et al, 1959]. In different individuals, the arthrogryposis sequence may be due to different causes. A sequence, therefore, may be a (Hippocratic) syndrome—a constellation of signs that, in different individuals, stems from the same pathogenesis but with different causes (eg, the nephrotic syndrome).

The word *paradrome* is suitable to designate any constellation of defects derived from a disturbance in a single developmental field; it encompasses three classes of the above classification: the monotopic field defects, the polytopic field defects, and the sequences. Paradromes can, therefore, be subdivided into three types: monotopic, polytopic, and sequential. Thus, paradrome is the shortest and most suitable term for designating the results of developmental field disturbances (eg, Hanhart's paradrome). Etymologically, this term means "a running side by side," whereas syndrome means "a running together."

3) Action of a single cause on more than one developmental field leading to a constellation of defects that are derived from two or more dysmorphogenetic processes. These constellations of defects are called malformation syndromes (or only syndromes, for short). Thus a syndrome may be defined as a pattern of multiple malformations presumed to derive from the action of a single cause on two or more developmental fields. Examples are Down's syndrome, Ellis-van Creveld's syndrome, thalidomide syndrome, Smith-Lemli-Opitz's syndrome. Syndromes may be due either to intrinsic defects (for example, Ellis-van Creveld's) or to a disruptive process (thalidomide syndrome).

4) Occurrence of two or more manifestations in the same individual due to linkage, linkage disequilibrium, predisposition, etc [Opitz, 1979]. This phenomenon should be called "association." Blood groups and diseases are good examples of association. Normal traits are also found associated in populations with racial stratification.

5) Action of different causes, present by coincidence in the same individual, on different developmental fields leading to different, nonpathogenetically related defects (eg, homozygosity for two different autosomal recessive traits). These events have been called concurrence, random association, chance syndrome [Cohen, 1977], and coincidental syndromy [Opitz, 1979]. To avoid confusion, we propose to name them concurrence or syntropy [see Opitz, 1979].

Concurrences are unclassifiable, since any two or more different defects may, by chance, be present in the same individual. Examples are epidermolysis bullosa dystrophica and Pelger-Huët's anomaly [Cat et al, 1967], aplasia or hypoplasia of upper lateral incisors and anonychia totalis [Freire-Maia and Pinheiro, 1979].

When, for unknown reasons, some defects are seen together at frequencies higher than those predicted by the simple product of their respective frequencies, this group may also be called association. An example is the VATER association (Vertebral defects, Anal atresia, T-E fistula with esophageal atresia, Renal defects, and Radial limb deficiency). With future advances in the knowledge of cause and pathogenesis, this kind of "association" may later turn out to be a syndrome, a paradrome, etc. Contrary to the well-established meaning of the word, the recommendations of an international working group [Spranger et al, 1982] state that only this type of constellation of signs should be called an association.

The use of the word association for these types of conditions is very restrictive, since very few of them have yet received this provisional label. Before the publication of writings on associations, it is useful to have private and provisional names for these apparently nonrandom occurrences of several defects in the same patient. However, experience shows that after being largely used in journals and books, names that should be provisional tend to persist much longer than necessary. We should not, therefore, use a "public" term for a provisional concept.

VATER association should be called VATER syndrome on the basis of the fact that a number of syndromes exist for which the evidence of being a *true* syndrome is as good as that for VATER. "...the term [syndrome] may be applied *unintentionally* to the complex manifestations of a single (perhaps polytopic) developmental field defect *as yet not recognized as such*" [Opitz, 1981b, page 114; our italics]. Fortunately, the nomenclature is not changing as rapidly as might be expected from the large number of recent suggestions to change it. So, let us not use another name (association) for situations in which syndrome, paradrome, etc, could be much better (provisional) labels. As pointed out by Opitz [1979], the word association has a "well-established meaning" (page 97); it should not be used with a different meaning or for an expected short time.

Marfan's "syndrome" has been proven to be a constellation of disturbances related to a basic defect in collagen. Since it is now known that the entire Marfan's clinical picture derives from a connective tissue dysplasia, the designation Marfan's dysplasia would be more correct. This is an example that shows that we do not need the provisional term (albeit against a well-established practice) "association" to designate conditions of as-yet unknown cause and pathogenesis. We first must use the designation we consider the most appropriate at the time (Marfan's syndrome) and, if it is later shown that the term is incorrect, change it to the correct one (Marfan's dysplasia).

SYNDROMES

As mentioned by Warkany [1971], malformation syndromes may be classified according to different principles: 1) localization and type of the defects, 2) histological or embryonic origin, 3) biochemical basis. The preference for one or another depends very much on the scientific outlook of the specialist. A clinician, an embryologist, or a biochemist may see the same entity from different perspectives and prefer to call it by different names. Terms such as limb deficiency, ectodermal dysplasia, and mucopolysaccharidosis reflect these three different outlooks. Since limb deficiency and ectodermal dysplasia may represent only parts of complex syndromes, the classification of syndromes under these labels should be used with great caution so as not to become misleading for the nonspecialist. For example, a condition should be labeled a dysplasia only if it is characterized by clearly dysplastic signs or by signs that may be related to a basic dysplastic defect; if these signs occur together with one or more malformations without being pathogenetically related to a dysplasia, the condition should be named a syndrome. Thus the condition would be called a (pure) dysplasia in the first case and a dysplasia/malformation syndrome in the second. However, this is not the opinion of the International Working Group [Spranger et al, 1982]; although requiring the existence of multiple developmental field defects to designate both a malformation syndrome and disruption syndrome, this group only required that a dysplasia be polytopic to be called a dysplasia syndrome.

Classifications based on biochemical data are assumed to be more "natural" than those based on anatomical, histological, or embryological similarities. However, as mentioned by Warkany [1971], "even such classifications [based on biochemical similarities] are sometimes short-lived when the chemical determinations shift from urine to blood or tissues or when clinical and chemical analyses are refined" (page 366). Biochemical traits are assumed to be nearer the first gene products than anatomical defects, but it is clear that there are biochemical hierarchies in the same manner as there are developmental hierarchies. Anyway, we are inclined to assume that classifications based on biochemical disturbances may reveal closer pathogenetic similarities than those merely based on, say, the localization and type of malformation.

Since many syndromes have been seen only once by a given investigator or group of investigators, it may happen that different researchers may have seen the same syndrome. If their data are not published, the syndrome will remain unknown to the scientific world though very well-known by isolated specialists. Our opinion is, therefore, that all apparently unique constellations of signs should be published, even if in short case reports, to call attention to the condition and to "flush out" similar cases. It is well known that many reports may follow the publication of an initial clinical description, thus contributing to a more complete delineation of the condition.

There is a widespread need among physicians to make a diagnosis if a syndrome has been seen in one or two patients. If no diagnosis can be made, all the precious information will be filed under "unknown syndrome." What could have been the birth of a "new" syndrome—if a paper had been written on it—frequently turns out to be a sad stillbirth. At the risk of publishing data on a syndrome so rare that little scientific interest may exist in its publication, we maintain that it is still better to run this risk and give other scientists an opportunity to continue the work and to cooperate in the delineation of a "new" condition than to be the father of a scientific stillbirth.*

These facts lead us to refer, for example, to Laurence-Moon-Biedl's syndrome that, before 1937, was thought to consist of a well-defined pentad: obesity, hypogenitalism, retinitis pigmentosa, polydactyly, and mental retardation. However, some patients who were supposed to have this syndrome, including some of those originally described by Laurence and Moon and by Biedl, did not have all the cardinal signs, ie, the full picture of the syndrome, but only some of them ("incomplete cases"). As time passed, the number of signs for the "complete" syndrome increased, variations were described among them and some investigators went as far as to state that the patients described by Laurence and Moon did not present Laurence-Moon-Biedl's syndrome! [For references, see Warkany, 1971]. Finally, recent investigation showed that this formerly chaotic complex is composed of at least two different syndromes—Laurence-Moon's syndrome and Bardet-Biedl's syndrome [for references, see McK 24580 and McK 20990, respectively].

Another situation that may be mentioned involves the ectrodactyly-ectodermal dysplasia-clefting and odontotrichomelic syndromes, grouped into a single entity by some authors and considered to be two different conditions by others. Obviously they share some similarities, and it is possible that other conditions, such as those described by Rapp and Hodgkin [1968], Bowen and Armstrong [1976], and Hay and Wells [1976], which resemble each other and the two conditions in question, form a nosologic group with more profound and intimate clinical similarities than many other nosologic groups [Pinheiro and Freire-Maia, 1980].

All of this discussion is related to the problem of the clinical delineation and classification of malformation syndromes. Patients may show partial

---

*The *Journal of Clinical Dysmorphology* (formerly *Syndrome Identification*), published by the March of Dimes Birth Defects Foundation (1275 Mamaroneck Ave., White Plains, NY 10605), is an appropriate vehicle for such new observations; in its Birth Defects Information Service (BDIS), 171 Harrison Ave., Box 403, Boston, MA 02110, the foundation offers its computer capacity to those who wish to match information on previously unidentified syndromes.

expression of a given cause (eg, an autosomal dominant or recessive gene) and, even in the same family, they may show different or overlapping constellations of signs. Thus, it may be difficult to determine whether similar constellations seen in different families represent the same syndrome or different ones. The study of additional families with cases identical to those found in the previous ones may lead to the hypothesis that we are really dealing with a single entity. On the other hand, if different kindreds can be sorted into two or more groups on a clinical basis, then it is reasonable to assume that we are dealing with two or more different conditions. The task of clearly delineating and differentiating very similar syndromes may sometimes require the patient collection and analysis of pedigrees, a time-consuming task. However, if two similar conditions are shown from the very beginning to be due to different causes, this information is sufficient to permit their separation into two entities, however similar they may look.

Since, in general, no obligatory or pathognomonic signs occur in any syndrome and diagnoses are made on the basis of the combination of a number of clinical signs, overlap among syndromes is the rule and the same syndrome may be described two or more times with different labels or may be seen just once and disappear from the scientific literature. The warning given by Warkany [1971] against "syndrome lovers" who may become "immortal" if ignorance persists regarding the conditions that they described and that received their names as eponym is rather pertinent. Attention is called to the conditions that now could be called Emed-Edwards' syndrome, McBride-Wiedemann-Lenz's syndrome, Gregg's syndrome, and Warkany's syndrome but are respectively known as trisomy 18 syndrome, thalidomide syndrome, maternal rubella syndrome, and aminopterin syndrome. Eponymic designation—valuable to honor distinguished scientists—should always be replaced with scientific terms, the best of which are obviously those connected with cause and pathogenesis. However, we would like to call attention to the fact that a number of syndromes whose cause is well known (for example, those due to simple Mendelian mechanisms) continue to have eponymic designations in the absence of other appropriate terms. Synonymy is common is syndromology, and when a given name begins to be used preferentially (for whatever reason), the tendency is to retain it. Thus, for example, the eponym Ellis-van Creveld is used more frequently than chondroectodermal dysplasia. The same applies to Poland's "syndrome," widely investigated but of unknown cause.

Unfortunately, the above meaning of such a basic concept as syndrome is not accepted by all specialists. For Cohen [1977], a syndrome is defined "simply as two or more abnormalities in the same individual" (page 103). [See also Cohen, 1976.] Thus the concurrence of epidermolysis bullosa dystrophica and Pelger-Huët anomaly in the same patient [Cat et al, 1967] would be called a syndrome, as would be the presence of bilateral brachy-dactyly of the second finger in an albino [Freire-Maia et al, 1978]. In several patients, we described the presence of recessive anonychia totalis with dominant hypoplasia of the upper lateral incisors [Freire-Maia and Pinheiro, 1979]. These three situations would be called "chance syndromes," ie, "the fortuitous occurrence of two or more abnormalities in the same individual" (page 111) by Cohen [1977]. This idea, if generally accepted, would certainly introduce another source of confusion in a difficult field. As mentioned, a suitable term for them is syntropy or concurrence.

For Beighton [1978], both Roberts' syndrome and acheiropodia [A. Freire-Maia, 1968, 1975; A. Freire-Maia et al, 1975, 1978] are "limb reduction syndromes." The first is characterized by variable degrees of limb deficiency (from minor deficiencies of segments to phocomelia) plus unusual facial appearance, hypotrichosis, different visceral abnormalities, severe growth retardation, etc. Compare this constellation of signs proba-bly representing disturbances of different embryological primordia with the clinical simplicity of acheiropodia—a congenital "amputation" above the elbows and below the knees (a tetra-hemimelia). Acheiropodia (wrongly named acheiria by Rimoin et al [1979]) should be called a polytopic para-drome or a polytopic field defect, *not* a syndrome. The same applies to the conditions we described more than two decades ago that includes variable bone aplasias and hypoplasias of the upper limbs [Freire-Maia et al, 1959].

We also had reported the presence of ectrodactyly, polydactyly, syndac-tyly, and flexion deformity in a girl born to consanguineous parents, and called this condition a syndrome [Fonseca and Freire-Maia, 1970]. This was a mistake, in spite of the fact that, at the time, the term syndrome did not have the more precise meaning it has gained more recently.

The process of defining syndromes may go through three stages [Opitz et al, 1969]:

1) The physical (or clinical) examination stage, in which a group of malformations detected in one patient is assumed to derive from the action of a single cause on two or more developmental fields.

2) The formal genesis (or pathogenetic) stage, when different patients found to have the same clinical picture are assumed to owe their complex

picture to similar dysmorphogenetic processes. However, the "same" clinical picture may have different causes in different patients and may represent an array of syndromes.

3) The causal genesis (or causal) stage, reached when similar constellations of signs observed in different patients are shown to have the same cause. In this case, the group of patients with a similar clinical picture presents a unique syndrome, *not* a heterogeneous group of similar syndromes. The highest position in the third stage is occupied by the syndromes for which we know not only the cause but also the biochemical agent (or pathway) that triggers the clinical picture. Hurler's syndrome is a paradigm in this field: It is due to an autosomal recessive gene that codes for a protein with defective α-iduronidase activity. Mucopolysaccharidoses, which a few years back formed a rather complex array of similar and poorly differentiated syndromes, now represent a good example of how heterogeneity can be resolved both causally and pathogenetically. The basic biochemical difference among them is so well defined that a mixed culture of cells from patients with two different types of these disorders leads to mutual biochemical correction. However, causal considerations were fundamental for separating them at the beginning (for example, Hunter's syndrome is due to an X-linked gene).

Some conditions, observed thus far in one patient only, remain in the first stage, awaiting the detection of similar cases. And they may remain there forever. Many others, observed in several patients, attain the second stage and stop there. Others, after discovery of the cause, attain the third and final stage. However, it is interesting to note that syndromes with the same—although unknown—cause (such as those observed in different members of the same sibship) belong, in a manner of speaking, to the bottom rung of the third stage.

A few examples of conditions now belonging to each of these three categories are given below:

Stage 1. Trichoodontoonychodermal syndrome, ectodermal dysplasia with severe mental retardation, Hayden's syndrome, arthrogryposis and ectodermal dysplasia, regional ectodermal dysplasia with total bilateral cleft, oculodentodigital syndrome II.

Stage 2. Rubinstein-Taybi's syndrome, Noonan's syndrome, Prader-Labhart-Willi's syndrome, Hallermann-Streiff's syndrome, alopecia universalis-onychodystrophy-total vitiligo.

Stage 3. Down's syndrome, cri-du-chat syndrome, Ellis-van Creveld's syndrome, hemophilia A, odontotrichomelic syndrome, Christ-Siemens-Touraine's syndrome, trichodentoosseous syndrome, Fischer-Jacobsen-Clouston's syndrome, trichoodontoonychial dysplasia.

Investigations on just one family may advance the process of defining a syndrome directly to the third stage if several relatives are affected and the same cause is detected or at least suspected in all patients. Many examples could be given of such a jump from the first to the final stage of clinical and etiological definition of a condition. The odontotrichomelic syndrome, for example, has been observed in four individuals from the same sibship and the practically identical clinical picture exhibited by all of them is likely to be due to the same (unknown) cause (autosomal recessive inheritance?). A second example, with an extremely high likelihood of autosomal recessive inheritance, is that of trichoodontoonychial dysplasia, also detected in four patients from a single sibship. However, it is important to keep in mind that stage 3 has several steps and that conditions such as the odontotrichomelic syndrome and trichoodontoonychial dysplasia are on the bottom step, whereas Hurler's and Hunter's syndromes are on the last and topmost step.

DYSPLASIAS

Following the conceptual line stressed by Herrmann and Opitz [1974] and Opitz [1979, 1981a,b], the International Working Group very wisely stated that "a dysplasia is an abnormal organization of cells into tissue(s) and its morphologic result(s)" [Spranger et al, 1982; page 162]. A dysplasia is, therefore, both the process and the consequence of dyshistogenesis, a developmental disturbance of tissue, hence of histological structure. However, this clear concept is contradicted in the same paper when, in the dendrogram representing the "nature of problem" (its Fig. 4), it is suggested that a polytopic dysplasia should be called a dysplasia syndrome. We cannot agree with this suggestion; if accepted, it would be another source of confusion. If a condition known as a syndrome is shown to have its constellation of signs originated from dyshistogenesis (not from dysmorphogenesis), the name syndrome should be abandoned and replaced with dysplasia. In recent years, osteogenesis imperfecta and Marfan's "syndrome" were shown to be due to defects in connective tissue [for references, see Spranger et al, 1982]. Therefore, these two conditions should be called dysplasias, *not* syndromes.

Following Spranger et al [1982], we would like to recall that dysplasia may have a metabolic (eg, the mucopolysaccharidoses) or a nonmetabolic nature (eg, giant cavernous hemangioma); involve only one germ layer (the unidysplasias such as the pure ectodermal dysplasias; eg, trichodental dysplasia) or several germ layers (eg, tuberous sclerosis); be generalized (eg, skeletal dysplasias) or localized (eg, presacral teratoma); single (eg, the solitary acoustic neuroma) or multiple (eg, multiple neurofibromata); benign (encephalotrigeminal angiomatosis) or premalignant (eg, colonic polyposis); permanent (eg, the signs of the ectodermal dysplasias) or evanescent (eg, cavernous hemangioma); congenital (most of the hair and nail signs in ectodermal dysplasias) or postnatal in appearance (eg, a testicular teratoma). Some ectodermal signs (for example, those related to the teeth) cannot be detected at birth, for obvious reasons.

Etymologically, the word dysplasia means abnormal growth, which is different from dysgenesis (abnormal formation). In this book, as mentioned, the term dysplasia is being used with the meaning of abnormal growth of tissue [Hermann and Opitz, 1974]. The French word *histodysplasie* would, therefore, be redundant in our context.

The word dysplasia is also used to designate conditions with signs restricted to connective tissue of bone and cartilage, muscle, etc, such as skeletal dysplasia, chondrodysplasia, arthromyodysplasia, pseudoachondroplastic dysplasia, and asphyxiating thoracic dysplasia. In other cases, the term dysgenesis seems preferable, such as in gonadal dysgenesis, familial ovarian dysgenesis, or thyroid dysgenesis. However, sometimes both terms are used as synonyms: dysgenesis of inner ear/inner ear dysplasia; mesodermal dysgenesis of iris and cornea/posterior marginal dysplasia of cornea.

The word dystrophy (etymologically, abnormal nourishment) is generally restricted to conditions that develop some time after birth, such as myotonic dystrophy or progressive muscular dystrophies. However, this word is sometimes also used as a synonym of dysplasia (eg, cranio-carpotarsal dysplasia/carpotarsal and cranial dystrophy; chondroectodermal dysplasia/polydactyly and chondrodystrophy; hidrotic ectodermal dysplasia/hereditary ectodermal dystrophy.

Dysostosis (abnormal development of bones) is also used, sometimes, as a synonym of dysplasia (of bones). Examples are hereditary metaphyseal dysostosis/metaphyseal chondrodysplasia; orodigitofacial dysostosis/dysplasia linguofacialis; unilateral mandibulofacial dysostosis/oculo-vertebral dysplasia; fibula dysplasia/fibula dysostosis.

Even though dysplasia should be applied to tissues, dysgenesis to organs, dysostosis to bones, dystrophy to conditions appearing after birth, etc, these words are sometimes used loosely as synonyms, compounding terminological confusion in this area.

As regards aplasia and hypoplasia, it is well known that they are sometimes used with the same meanings of agenesis and hypogenesis, respectively. Examples are radial aplasia or agenesis, thumb hypoplasia or hypogenesis. However, Spranger et al [1982] suggest that these and other terms should be used with precise meanings. Since most of their suggestions for definitions appear sound, they are repeated here for the sake of completeness. Agenesis: absence of a region due to an absent primordium; aplasia: absence of a region due to failure of the primordium to develop; atrophy: decrease of size of a normally developed region as a consequence of a decrease in cell size and/or number; hypoplasia and hyperplasia: underdevelopment and overdevelopment of an organism, organ or tissue resulting from a decreased or increased number of cells, respectively; hypotrophy or hypertrophy: decrease and increase in the size of cells, tissues or organs, respectively. This definition is too short and unclear. The authors probably wish to say that the underdevelopment and the overdevelopment of tissues and organs are due to a decrease or an increase in the *size* of cells, since, as defined above, when this happens as a consequence of a decrease or increase in the *number* of cells, the conditions should be called hypoplasia and hyperplasia, respectively.

Some of the above terms are covered by the widest concept of dysplasia. For instance, enamel hypoplasia is a dysplasia.

Dysplasias may be confined to a single point or to multiple points of tissue from the same germ layer or affect tissues from more than one layer. Pure dysplasias have been classified as follows, according to the number of germ layers involved and to the number of dysplastic signs produced [Herrmann et al, 1977]: Number of germ layers involved: 1) unidysplasias (from one germ layer), 2) multidysplasias (from two or three germ layers). Number of dysplastic signs: 1) monodysplasias (a single dysplastic sign), 2) polydysplasias (two or more dysplastic signs).

On the basis of this classification, alopecia universalis-onychodystrophy-total vitiligo, Naegeli-Franceschetti-Jadassohn's dysplasia, palmoplantar hyperkeratosis and alopecia, etc, are pure unipolydysplasias since they are characterized by multiple dysplastic signs in tissues apparently derived from the ectoderm only. On the other hand, ectodermal dysplasia of the head and hypohidrotic ectodermal dysplasia with hypothyroidism are multipolydysplasias because they include signs involving different germ layers. However, conditions such as Christ-Siemens-Touraine's syndrome, Ellis-van Creveld's syndrome, Rothmund-Thomson's syndrome, Coffin-Siris' syndrome, and odontotrichomelic syndrome are dysplasia/malformation syndromes. Ellis-van Creveld's syndrome, for example, is a multipolydysplasia associated with malformations, such as polydactyly.

On a general view on malformations, syndromes and dysplasias, see Freire-Maia [1982].

ECTODERMAL DYSPLASIAS—A HISTORICAL APPROACH

The term hereditary ectodermal dysplasia was coined by Weech [1929] to substitute for other expressions such as dystrophy of the hair and nails, imperfect development of skin, hair, and teeth, and congenital ectodermal defect that had been used to designate a small group of conditions characterized wholly or in part by hypotrichosis, hypodontia, onychodysplasia, and "anhidrosis." According to Weech, the new designation applied to the three essential aspects of these conditions: 1) most of the disturbances affect tissues of ectodermal origin; 2) these disturbances are developmental; 3) heredity plays an important causal role.

The best known of these conditions was one (the "anhidrotic form") now known as the Christ-Siemens-Touraine's (CST) syndrome and also under the misleading term X-linked anhidrotic ectodermal dysplasia [Freire-Maia and Pinheiro, 1980], which is characterized by hypotrichosis, hypodontia, and severe hypohidrosis as well as by saddle nose, frontal bossing, unusual skin wrinkling, scanty eyebrows and lashes, protruding lips, and chronic rhinitis and pharyngitis associated with a decrease of sense of taste and smell.

This "form" was generally reported as due to a recessive sex-linked gene. Weech [1929] stated that "were no more to be said, it would seem that the inheritance data just given would be in themselves sufficient to separate the anhidrotic from other forms of ectodermal dysplasias" (page 785). From a clinical point of view, Weech cited several aspects "favouring such a division" in the teeth, the sudoriparous glands, the nails, etc. However, he then goes on to note that "full-blown" cases of the so-called anhidrotic form were at times found in females. He points out that this fact "does much to weaken the full acceptance of the sex-linked character of the inheritance in the anhidrotic group" (pages 785–786). On page 787, he confesses that "an adequate explanation for the occurrence of the anhidrotic syndrome in the female is not possible." Now we know that what he called the "anhidrotic syndrome" consists of at least two syndromes, one due to an X-linked gene (which may have mild to severe expression in heterozygous females)—the CST syndrome—and the other due to the homozygous state of an autosomal recessive gene—autosomal recessive hypohidrotic ectodermal dysplasia [Passarge et al, 1966; Gorlin et al, 1970; see also Passarge and Fries, 1977].

The first ectodermal dysplasia presented in the literature appears to be that of Danz [1792], who reported two Jewish boys with congenital atrichia and anodontia without any further comment.

The first cases of CST to be reported in the literature seem to be those of Thurnam [1848]: two severely affected male first cousins and their mildly affected maternal grandmother. A kindred studied by Wedderburn in 1838 was mentioned in 1875 by Charles Darwin in his book *The Variations of Animals and Plants under Domestication* [cf Darwin, 1880]. This condition was redescribed by Thadani [1921]: The affected males were bald, toothless, and had extreme intolerance to heat.

Subsequently, other investigators have described clinically similar cases in a large number of families, with consequent problems of heterogeneity [see Weech, 1929; Touraine, 1936, 1952; Upshaw and Montgomery, 1949; Franceschetti, 1953; Durham, 1960; Rosselli and Gulienetti, 1961; Greene, 1962; Montgomery, 1967; Rubin, 1967; Jablonski, 1969; Gorlin et al, 1970; Freire-Maia, 1971, 1973, 1977a; Settineri, 1974; Witkop et al, 1975; Bergsma, 1979; McKusick, 1983]. Cardinal signs, associated signs, and general symptoms have been recognized; severe, moderately severe, and mild states have been identified; "incomplete syndromes" and "variants" have been described. The cardinal signs and some of the most common associated signs of ectodermal origin justified the general expression covering the group. However, the high clinical and causal heterogeneity of the group showed that we were dealing with a great number of different conditions rather than a few highly variable conditions. With time, this view became well established. As more and widely varying associated signs (many of them affecting tissues not of ectodermal origin) were identified and some of the cardinal ones were found not to be obligatory, the complexity of the nosologic group grew larger and larger. Touraine [1936] suggested the expression ectodermal polydysplasia in place of the term anhidrotic form to call attention to the extensive polysymptomatology involved and, during the last decade, it became generally accepted that this "form" was only one among a number of other conditions, each equally entitled to be called an ectodermal dysplasia.

Entities were first classified into two groups—anhidrotic and hidrotic. However, since hypotrichosis, dental defects, or onychodysplasia could be present or absent in each entity, even the expression ectodermal dysplasia started to lose its earlier, clear significance. How many ectodermal signs— and which ones—must a condition have to be classified as an ectodermal dysplasia? Should anyone with *one* sign affecting hair or teeth or epidermis be designated an ectodermal dysplasia? If so, the number of ectodermal dysplasias would be so large that the expression would entirely lose its practical value. Should the condition then have two, three, four ectodermal signs? But which combinations of signs should be accepted as criteria for

classification? In other words, what clinical criteria should define an ectodermal dysplasia? As regards sweating, there is euhidrosis (normal sweating), hypohidrosis, and hyperhidrosis. Dental defects are even more
variable and include anodontia, hypodontia, persistence of deciduous teeth,
delayed eruption, enamel hypoplasia, peg-shaped teeth, and supernumerary teeth, varying widely from one patient to another (with the same
condition) and from one condition to another. With respect to trichodysplasia, hypo- and hypertrichosis may occur in different degrees. The same
kind of variability is found for onychodysplasia. Weech [1929], for example, had already noted that the nail defect in the condition described and
reviewed by Jacobsen [1928] was never seen in "his" anhidrotic form.

What then is an ectodermal dysplasia? The answer varies from author
to author. It may be any one of the following:

1) One condition: The sex-linked recessive classical hypohidrotic form
(CST syndrome) [Alexander and Allen, 1965; Dominok and Rönisch,
1968; Bollaert and Wachholder, 1969; Agostinelli, 1970].

2) A group of two well-defined conditions: The CST syndrome and
Clouston's syndrome [Blattner, 1968; Shore, 1970; Wilbur, 1973; Gwinn
and Lee, 1974; Reddy et al, 1978], the former considered to be the
anhidrotic type and the latter the hidrotic type.

3) A group of two main (major) conditions—the anhidrotic and hidrotic
forms—with some heterogeneity [Clouston, 1939; Robinson et al, 1962;
Lowry et al, 1966; Samuelson, 1970; Machtens et al, 1972].

4) A group of a few conditions [Weech, 1929; Blassingille, 1959; Greene,
1962; Smith, 1969, 1970; Mochizuki et al, 1971; Solomon and Esterly,
1973]. A comparison of the lists of ectodermal dysplasias suggested by
these authors suffices to reveal the profound discrepancies that exist among
them. Since none of these authors proposed a definition of ectodermal
dysplasia, any condition with some signs of ectodermal origin could be
termed an ectodermal dysplasia.

5) A large group of conditions as defined by Freire-Maia [1971] (see
below).

These five concepts have been analyzed by Freire-Maia [1977a]. We
accept the fifth one, which is also accepted by other authors [Settineri,
1974; Witkop et al, 1975; Pinsky, 1975, 1977; Pinheiro, 1977, 1983; Ra-
pone-Gaidzinsky, 1978; Solomon and Keuer, 1980; Hazen et al, 1980;
Alves et al, 1980, 1981; Peterson-Falzone et al, 1981; Pinheiro et al, 1981b,
1983a, c; Smith, 1982; Freire-Maia and Pinheiro, 1983a; Pinheiro and
Freire-Maia, 1983].

A pathogenetic definition of ectodermal dysplasia is "a developmental
defect that, at the embryological level, affects the ectoderm." Clinically, it

is represented by a group of conditions with signs in tissues of ectodermal origin (pure ectodermal dysplasias) and generally also in structures of nonectodermal origin, as mentioned earlier. This means that the heading ectodermal dysplasias covers pure unidysplasias, pure multidysplasias, and malformation/dysplasia syndromes, according to the nomenclature proposed by Herrmann et al [1977]. Since the condition may vary in extent and severity according to the occurrence of the primary defect earlier or later during the embryological development [Greene, 1962], the same condition may show high clinical variability.

Some of the conditions that will be listed in this book as ectodermal dysplasias have been classified into four different groups by Smith [1970]: Hallerman-Streiff's ("unusually small stature with associated defects"); oculodentodigital syndrome ("oral-facial-digital associations of defects"); focal dermal hypoplasia, Rothmund-Thomson's, dyskeratosis congenita and incontinentia pigmenti ("hamartoses"); and Ellis-van Creveld's ("osteochondro-dysplasias"). As mentioned by Smith [1970], this is an arbitrary classification, as classifications of conditions sometimes are.

ECTODERMAL DYSPLASIAS—A TAXONOMIC APPROACH

Ectodermal dysplasias are defined as conditions with at least one of the signs detected in the first cases reported in the literature—trichodysplasia, dental defects, onychodysplasia, and dyshidrosis—plus at least one sign affecting another structure of ectodermal origin [Freire-Maia, 1971, 1973, 1977a, 1982]. This, of course, is an artificial definition formulated only with the intent of delimiting the field. Therefore, conditions affecting only, say, the skin and oral mucosa (ichthyosis congenita), which have been already classified among ectodermal dysplasias [Greene, 1962], will not be covered by our definition.

Our words should not be interpreted to mean that *any* defect affecting, say, the teeth, nails, hair or skin (for example, acrodermatitis enteropathica and congenital porphyria) will be called ectodermal dysplasia. Rather, it is implied that the clinical signs detected in those regions have derived from what is assumed to be a primary ectodermal defect.

As already mentioned, ectodermal dysplasias generally are not complexes of signs of a purely ectodermal origin. The old concept of "one layer disease" is no longer acceptable [Rosselli and Gulienetti, 1961] since many conditions are characterized by constellations of signs of multiple embryological origin. The original name remains only because of historical considerations and because the ectodermal defects are either more severe or more apparent than the others.

The process of naming is highly confusing in this field. Different eponyms side by side with different descriptive names for the same condition have created an extensive nomenclature. Jablonski [1969], Smith [1970], Witkop et al [1975], Gorlin et al [1976], and Freire-Maia [1977a], among others, should be consulted in case of doubt.

This book describes 108 clinical conditions with at least *two* of the four signs mentioned above (Group A), comprising a total of at least 117 ectodermal dysplasias, due to heterogeneity. They are classified according to the number assigned to these signs: trichodysplasia (1), dental defects (2), onychodysplasia (3), and dyshidrosis (4). The different subgroups may be called either by the combination of the above numbers (for instance, 1-2-3-4, 1-2-3) or by the combination of words of Greek origin referring to the structures showing the disturbances. Thus, 1-2-3-4 equals tricho-odonto-onycho-dyshidrotic; 1-2-3 equals tricho-odonto-onychic; 1-2-4 equals tricho-odonto-dyshidrotic; 1-3-4 equals tricho-onycho-dishydrotic; 2-3-4 equals odonto-onycho-dyshidrotic; 1-2 equals tricho-odontic; 1-3 equals tricho-onychic; 1-4 equals tricho-dyshidrotic; 2-3 equals odonto-onychic; 2-4 equals odonto-dyshidrotic; and 3-4 equals onycho-dyshidrotic. When referring to a specific condition, the word *dyshidrotic* may be replaced by *hypo-* or *hyperhidrotic*. For example, CST syndrome is a tricho-odonto-onycho-hypohidrotic condition.

Conditions with at least *one* of the four signs mentioned above plus at least one sign in another structure of ectodermal origin (labeled 5) are classified into another group (Group B). These conditions may be called either by the combination of the respective numbers or by the words of Greek origin referring to one of the four basic signs. Basan's syndrome, for example, belongs to the 3-5 subgroup or to the onychic subgroup; pili torti and deafness belongs to the 1-5 subgroup or to the trichic subgroup. The subgroups may also be called by the corresponding terms of any vernacular language. Examples would be hair-tooth-nail subgroup (for 1-2-3) or nail subgroup (for 3-5). Group B seems to be smaller than Group A, whose nosologic importance led us to devote this entire book to it.

Many conditions other than those mentioned here could be classified into the 1-2 subgroup, but convenience, based on a good knowledge of their pathogenesis, suggests that they should be classified into other groups. For example, Hurler's, Hunter's, and Sanfilippo's are to be maintained among the mucopolysaccharidoses; Treacher Collins' (mandibulofacial dysostosis) among the craniofacial dysostoses; etc. By the same token, homocystinuria (1-2), hyalinosis cutis et mucosae (1-2), progeria (1-2-3), Sjögren-Larsson's syndrome (1-2-4), and many others should preferably be maintained outside ectodermal dysplasias and classified into other more suitable groups. It may also happen that some of the conditions listed here as ectodermal dysplasia will later be moved to other groups. Therefore, our classification may be considered both incomplete and too crowded. However, this is purely a clinical-mnemonic classification, and we hope it will be changed as we progress in the knowledge of the pathogenesis of each condition.

As an example of a condition that is an ectodermal dysplasia sensu stricto but whose ectodermal signs are minor in relation to the other manifestations, Saethre-Chotzen's syndrome is described below in the same style we will use for the other conditions discussed in the next chapters. (AD, autosomal dominant.)

Saethre-Chotzen's Syndrome (McK 10140)

Synonyms. Chotzen's syndrome; pseudo-Crouzon's syndrome; acrocephaly with rudimentary syndactyly; acrocephalosyndactyly type III.

Hair. Low-set frontal and neck hairline.

Teeth. Enamel hypoplasia; malposition; hypodontia; peg-shaped.

Nails. Normal.

Sweat. No data.

Skin. Transpalmar crease.

Hearing. Impaired.

Eyes. Bilateral protrusion; imperfect eyeball mobility; optic atrophy; esotropia; exotropia; paleness of optic discs; occasional lacrimal duct abnormalities.

Face. Asymmetry; flattened nasofrontal angle; hypoplastic maxilla; widened nasal root; opened mouth; lateral deviation of the nasal septum; palpebral ptosis; antimongoloid slant of the palpebral fissures; dysmorphic auricles; mandibular prognathism; hypertelorism.

Psychomotor and growth development. Seizures; occasional mental retardation.

Limbs. Partial syndactyly of fingers; clinodactyly; brachydactyly; partially bifid distal phalanges of the great toes; hallux valgus; contractures at elbows and knees; webbing of some toes; wide great toes.

Other findings. Nonsymmetrical head of oxybrachycephalic configuration; craniosynostosis; lack of frontal sinus formation or pneumatization; hypoplasia of maxillary sinuses; highly arched palate; spinal anomalies; cleft palate; cryptorchidism; renal anomalies; congenital heart defect; deformed chest; winged costal arches.

Etiology. AD.

Comments. This condition is classified among the acrocephalosyndacty-lies since it combines craniofacial developmental anomalies (oxybrachyce-phaly) with abnormalities in the distal limbs (especially syndactyly).

Acrocephalosyndactylies have been classified into six types by McKusick [1975]: Apert's syndrome, Vogt's cephalodactyly (= Apert's syndrome?), Saethre-Chotzen's syndrome, Pfeiffer's syndrome, Summitt's syndrome, and Herrmann-Opitz's syndrome. These six acrocephalosyndactylies are included among the 14 "forms" of craniosynostoses, as mentioned by Cohen [1975]. The other eight conditions are Kleeblattschädel's "anom-aly" (see below), Carpenter's syndrome, Christian's syndrome, Baller-Gerold's syndrome, Lowry's syndrome, Gorlin-Chaudhry-Moss' syn-drome, Herrmann-Pallister-Opitz's syndrome, and Sakati-Nyhan-Tis-dale's syndrome. Kleeblattschädel's "anomaly" may occur alone or associated with other conditions, such as thanatophoric dwarfism, Crou-zon's syndrome, Apert's syndrome, Pfeiffer's syndrome, and Carpenter's syndrome.

Other references. Bartsocas et al [1970], Pantke et al [1975], Pruzansky et al [1975], Kopyść et al [1980]. (For the syndromes mentioned above, see Smith [1970] and Goodman and Gorlin [1977]).

Epidermolysis bullosa is a group of conditions characterized by bulla for-mation that differ from each other in terms of clinical features, severity, fre-quency, and etiology. Different numbers (from two to about ten) of "diseases," "forms," "syndromes," or "general forms" have been described under the heading of epidermolysis bullosa by different authors [for reviews, classifications, cases reports, etc., see, for example, Gorlin and Pindborg, 1964; Passarge, 1965; Davison, 1965; Bart et al, 1966; Schnyder, 1967; Gedde-Dahl, 1971; Howden and Oldenburg, 1972; Joensen, 1973; Gorlin et al, 1976; McKusick, 1979]. Some of the forms could be classified as ectoder-mal dysplasias, but since all of them form a well-delineated group, it seems preferable, contrary to our first impression [Freire-Maia, 1971], to leave the entire group untouched.

As regards incontinentia pigmenti, since most of the most apparent cardinal signs are related to the skin (patches of vesicles, inflammatory lesions, pigmented macules, hyperkeratotic areas, etc), it would seem preferable to classify it as outside of ectodermal dysplasias. However, for lack of a better group in which to include it, we will keep it in subgroup 1-2-3.

Our classification is also subject to revision both due to possible mistakes by the authors and to the fact that some conditions have not been fully investigated. In fact, because reliable methods have not always been applied to evaluate the sweating capacity of the patients, one or more conditions presented as euhidrotic may really be dyshidrotic and would then have to be moved to another group. The inverse may also occur: Since euhidrosis varies widely from person to person, patients sweating less than what is assumed to be normal may be erroneously described as hypohidrotic. The fact that some conditions have been described mainly in children or in a small number of adult patients, some of whom belong to the same family, may also account for too narrow a delineation. Finally, it should be mentioned that heterogeneity may be present in a larger number of conditions than those mentioned here. Anyway, the number of conditions is certainly larger than that reported in our final etiological overview.

SUMMARY

After reading this chapter, the reader will be correct in thinking that there is tremendous confusion in the realm of definitions and classifications of the conditions reviewed. Since many pages have been devoted to describing the chaos, we shall try to summarize in a few lines the basic definitions we accept.

Malformation

Morphologic defect of an organ, part of an organ, or a larger region resulting from the action of a single cause on a single developmental field. When so defined, it is a primary malformation (derived from an intrinsic defect of the primordium).

Disruption

Morphologic defect, as defined above, resulting from the action of an extrinsic cause on an originally normal primordium. Also called secondary malformation. A *deformity* is an abnormal form, shape, or position caused by mechanical forces.

Paradrome

Monotopic. Constellation of closely contiguous and pathogenetically correlated malformations derived from the action of a single cause on a single (monotopic) developmental field.

Polytopic. Constellation of distant but pathogenetically correlated malformations derived from the action of a single cause on a single (polytopic) developmental field.

Sequential. Constellation of malformations and/or other types of defects triggered by a single defect that derived from the action of a single cause on a developmental field. A *sequence* may be composed of malformations and/or dysplasias. It may also encompass functional disorders.

Syndrome

Constellation of malformations derived from disturbances produced by a single cause on two or more developmental fields (malformation or dysmorphic syndrome). A syndrome, therefore, is strictly defined here as a constellation of signs derived from a multiple-field dysmorphogenetic process. In this context, the expression "dysplasia syndrome" is as great an absurdity as malformation dysplasia. A condition combining both malformations and dysplasias can, however, be properly called a malformation/dysplasia syndrome.

A syndrome may represent a constellation of primary and/or secondary malformations. Therefore, the expressions "malformation syndrome" and "disruption syndrome" are justifiable.

Association

Occurrence in the same individual of two or more manifestations due to linkage, linkage disequilibrium, predisposition, race stratification, etc.

Concurrence (or syntropy)

Constellation of malformations and/or other types of defects present, by chance, in the same individual and thus derived from the action of two or more independent causes on different developmental fields.

Dysplasia

The redundant French word *histodysplasie* stresses the meaning of the word: It is both the process and the resulting clinical picture of a dyshistogenesis. A dysplasia, therefore, is related to a defect of tissue organization, whereas a primary malformation is the result of a more complex defect related to the formation of an organ, part of an organ or a larger region. Dysplasias may be classified into a number of groups but, in this summary, two categories seem to be especially important: pure dysplasias (constellations of signs that are entirely dysplastic in origin) and malformation/dysplasia syndromes (constellations of malformative and dysplastic signs). Conditions that look like syndromes but whose malformationlike signs really stem from dysplastic processes should, therefore, be called dysplasias.

Disease

This word should be restricted to conditions with progression and deterioration with time.

Examples

Malformation: cleft lip; disruption: rubella syndrome; monotopic paradrome: holoprosencephaly; polytopic paradrome: acheiropodia; sequential paradrome: myelomeningocele; syndrome: Down's syndrome; malformation/dysplasia syndrome: Ellis-van Creveld's syndrome; association: HLA-B27 and ankylosing spondylitis; concurrence (or syntropy): Pelger-Huët's anomaly plus epidermolysis bullosa; dysplasia: trichodental dysplasia; disease: Tay-Sachs' disease.

3. Subgroup 1-2-3-4

1. CHRIST-SIEMENS-TOURAINE'S (CST) SYNDROME
(McK 30510)

Synonyms. Hypohidrotic X-linked ectodermal dysplasia; sex-linked ectodermal dysplasia; anhidrotic ectodermal dysplasia; congenital anhidrotic ectodermal dysplasia; congenital ectodermal defect; Christ-Siemens-Weech's syndrome; Siemens-Weech's syndrome; ectodermal polydysplasia; anhidrosis-hypotrichosis-anodontia syndrome; anhidrosis hypotrichotica sexoligata; anhidrotic hereditary ectodermal dysplasia; hereditary ectodermal dysplasia; Jacquet's syndrome; Siemens' dermatosis; Siemens' syndrome; Weech's syndrome; etc.

Hair. Fine and dry; hypochromic; hypotrichosis of scalp and body; absent or scanty eyebrows and lashes; moustache and beard generally normal.

Teeth. Hypodontia; peg-shaped incisors and/or canines; persistence of deciduous teeth; delayed eruption; occasional anodontia.

Nails. Generally normal; sometimes dystrophic or absent at birth and/or fragile and brittle with incomplete development and celonychia.

Sweat. Hypohidrosis with or without hyperthermia; reduced response to pilocarpine by iontophoresis and heat; absence or decreased number of epidermal ridge sweat pores.

Skin. Thin, smooth, and dry due to hypoplasia or absence of sebaceous glands; occasional pigmentation and dermatoglyphic changes; absent or supernumerary nipples and areolae.

Hearing. Occasional conductive loss.

Eyes. Photophobia; decreased function of the lacrimal glands; aplasia or hypoplasia of lacrimal ducts.

Face. Highly characteristic in persons who are severely affected (generally males) with thick and prominent lips, depressed nasal bridge (saddle nose), frontal bossing, hypoplasia of the maxilla, wrinkles beneath the eyes or around the eyes, nose, and mouth, and minor alterations of the auricles. The periorbital region is often more darkly pigmented than the rest of the body.

Psychomotor and growth development. See *Comments.*

Limbs. See *Comments.*

Other findings. Atrophic rhinitis (sometimes associated with ozena and epistaxis); otitis media; decreased sense of taste and/or smell; atrophied mucous glands of the upper respiratory tract (leading to an increased susceptibility to infection); respiratory difficulties; chronic pharyngitis and laryngitis (with dysphonia and hoarseness); aplasia or hypoplasia of the mammary glands.

Etiology. X-linked recessive (XR); roughly 70% of the heterozygous women have partial or mild manifestations (minor form) of the syndrome (hypotrichosis, hypodontia, microdontia, hypohidrosis, patchy mosaic distribution of body hair and of sweat pores, etc) [Pinheiro and Freire-Maia, 1977, 1979c; Nakata et al, 1980; Pinheiro et al, 1981a; Sofaer, 1981a, b; Freire-Maia and Pinheiro, 1982a, b].

Comments. Thurnam [1848] seems to have been the first author to describe CST. Danz [1792] reported two Jewish men without hair and teeth, but did not give any further information about them. Wedderburn is often quoted as having described a Hindu family in 1838*, but his description did not appear until 1875, when Darwin quoted it in his book *The Variations of Animals and Plants Under Domestication* [cf Darwin, 1880].

Finger and toe defects, short stature, and mental retardation have also been described in a few patients but it is doubtful that these are signs of the condition.

More than 350 affected men have been reported. However, contrary to general opinion, affected women outnumber affected men [Pinheiro and Freire-Maia, 1979b].

Bernard et al [1963] reported a kindred with six severely affected women in three generations with a CST-like syndrome, severe retardation of

---

*References with the indication of the year preceded by "in" mean that the original paper or book was not read. Detailed references on these publications may be seen in other reviews as well as in general books, such as Durham [1960], Gorlin and Pindborg [1964], Rubin [1967], Montgomery [1967], Jablonski [1969], Nelson et al [1969], Smith [1970, 1982], Holmes et al [1972], Solomon and Esterly [1973], Gorlin et al [1976], Stewart and Prescott [1976], Goodman and Gorlin [1977], Beighton [1978], Salmon and Lindenbaum [1978], Wiedemann et al [1978], Bergsma [1979], Der Kaloustian and Kurban [1979], McKusick [1983].

psychomotor development, and early death. They suggested that CST may be due to different alleles with different degrees of expressivity and penetrance. Their pedigree is compatible with both X-linked and autosomal dominant inheritance.

There is interfamily clinical heterogeneity among both men and women, all situations being compatible with X-linked inheritance. Different alleles at the same locus or two or more X-linked loci may be involved in what may really be a small group of similar conditions.

Other references. Weech [1929], Hutt [1935], Touraine [1936], Felsher [1944], Upshaw and Montgomery [1949], Perabo et al [1956], De Jager

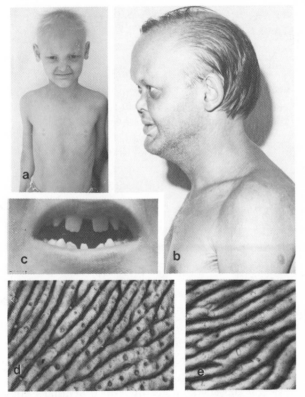

Fig. 3-1.   Christ-Siemens-Touraine's syndrome. a. Hypotrichosis, hypochromic hair, abnormal auricles, and supernumerary nipples. b. Hypotrichosis, frontal bossing, saddle nose, protruding lips, and normal beard and moustache. c. Hypodontia, microdontia, peg-shaped and widely spaced teeth in a carrier. d. Mosaic patchy distribution of sweat pores in a carrier. e. Absence of sweat pores in an affected man. a–c, unpublished photos; d,e, Pinheiro and Freire-Maia [1979a].

[1965], Alexander and Allen [1965], Kerr et al [1966], Lowry et al [1966], Blattner [1968], Dominok and Rönisch [1968], Frias and Smith [1968], Bollaert and Wachholder [1969], Olinsky and Thomsom [1970], Samuelson [1970], Shore [1970], Verbov [1970], Buhr and Schuster [1971], Crump and Danks [1971], Mochizuki et al [1971], Machtens et al [1972], Passarge and Fries [1973], Gwinn and Lee [1974], Familusi et al [1975], Messow et al [1977], Pinheiro [1977], Reddy et al [1978], Pinheiro and Freire-Maia [1979a], Freire-Maia and Pinheiro [1980], Chautard-Freire-Maia et al [1981], Kleinebrecht et al [1981], Airenne [1981].

2. AUTOSOMAL RECESSIVE HYPOHIDROTIC ECTODERMAL DYSPLASIA (McK 22490)

Synonyms. None.

Hair. Sparse, fuzzy, lightly pigmented scalp hair; absent or scanty eyebrows, lashes and body hair.

Teeth. Hypodontia; anodontia; conical teeth.

Nails. Generally normal; sometimes hypoplastic.

Sweat. Hypohidrosis with hyperthermia.

Skin. Thin, smooth, dry and hypoplastic.

Hearing. See *Comments.*

Eyes. Photophobia; hypoplasia of lacrimal ducts; decreased function of the lacrimal glands.

Face. Highly characteristic: saddle nose, thick and protruding lips, frontal bossing, and prominent auricles. The skin around the orbits is darker and wrinkled. See *Comments.*

Psychomotor and growth development. Occasionally low intelligence level (?).

Limbs. Normal.

Other findings. Chronic rhinitis (with associated ozena); frequent respiratory infections.

Etiology. Autosomal recessive (AR).

Comments. Passarge et al [1966] described three sisters born to consanguineous parents and their three first cousins (two boys and one girl) also from a consanguineous union. Gorlin et al [1970] referred to another affected girl and reviewed the literature, documenting a number of cases of this syndrome. Passarge and Fries [1977] reexamined the three sisters referred to earlier and detected severe midface hypoplasia. They are inclined to consider this syndrome as distinct from the autosomal recessive hypohidrotic ectodermal dysplasia mentioned by other authors. Sensorineural hearing loss and hypertelorism were only cited in the 1966 paper.

Other references. Hartwell et al [1965], Crump and Danks [1971], Bartlett et al [1972], Kratzsch [1972].

3. FOCAL DERMAL HYPOPLASIA (FDH) SYNDROME (McK 30560)

Synonyms. Goltz-Gorlin's syndrome; ectodermal and mesodermal dysplasia with osseous involvement; congenital ectodermal and mesodermal dysplasia; combined mesoectodermal dysplasia; focal dermato-phalangeal dysplasia; Goltz's syndrome; focal dermal dysplasia syndrome.

Hair. Hypotrichosis in focal areas of scalp and pubis.

Teeth. Hypodontia; microdontia; enamel hypoplasia; delayed eruption; irregular placement.

Nails. Thin, spooned, narrow, grooved, hypopigmented, or absent.

Sweat. Hypohidrosis; hyperhidrosis (especially of palms and soles).

Skin. Absence of the skin from various parts at birth; areas of underdevelopment and thinness; linear hypo- or hyperpigmentation; telangiectasia; herniation of subcutaneous fat; multiple papillomas of mucous membranes of periorificial skin; follicular hyperkeratotic papules; angiofibromatous nodules around lips and anus; palmoplantar hyperkeratosis; occasional dermatoglyphic changes.

Hearing. Occasional loss (sensorineural and conductive types).

Eyes. Colobomas; microphthalmia; anophthalmia; strabismus; nystagmus; irregularity of pupils; clouding of cornea or vitreous; blue sclerae.

Face. Lip papillomas; asymmetric development; malformed auricles; asymmetry and notching of the alae nasi; pointed chin; triangular face; hypertelorism.

Psychomotor and growth development. Mental retardation; short stature.

Limbs. Syndactyly; polydactyly; olygodactyly; adactyly; brachydactyly; clinodactyly; hypoplasia of the clavicles.

Other findings. Hypoplasia of the external genitalia; umbilical and/or inguinal hernia; vertebral anomalies (scoliosis, spina bifida, etc); highly arched palate; gum papillomas; small breasts.

Etiology. X-linked dominant (XD)? Autosomal dominant (AD)?. See *Comments.*

Comments. About 90% of the patients are women. With the exception of two kindreds, all cases reviewed by Ginsburg et al [1970] were sporadic (in one pedigree the syndrome was transmitted along four generations from mother to daughter; Gorlin et al, 1963). Ruiz-Maldonado et al [1974] described five sisters and their mother; among 16 Mexican cases thus far studied, 14 are women.

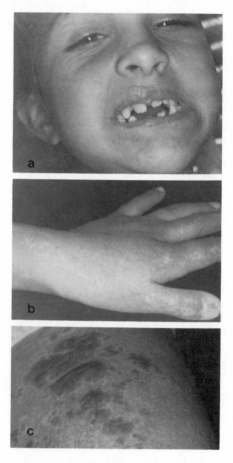

Fig. 3-2.   Focal dermal hypoplasia syndrome. a. Dental defects, asymmetric face, assy-metry of alae nasi, pointed chin, skin and lip lesions. b. Syndactyly between third and fourth fingers, linear hypopigmented areas, dystrophic nails. c. Skin lesions. Courtesy of Drs. Aida S. Libis, I.O. Pinto, and Judith Viégas, Pelotas, RS, Brazil.

Goltz et al [1962] suggested an X dominant gene, generally lethal in the men and reducing fertility in the women, as the cause of this condition. An alternative hypothesis is that of an AD gene. The same hypotheses have been suggested for incontinentia pigmenti. We think the XD hypoth-esis is the most probable of them. This problem will be discussed in Chapter 16.

Other references. Goltz et al [1970], Ishibashi and Kurihara [1972], Zer-gollern et al [1974], Toro-Sola et al [1975], Gorlin et al [1976], Libis et al [1982].

4. XERODERMA-TALIPES-ENAMEL DEFECT (McK: not listed)

Synonym. XTE syndrome.

Hair. Course and dry; slow-growing; hypotrichosis; no lashes on lower lids.

Teeth. Poorly formed; yellow enamel.

Nails. Deformed on the fingers and/or toes.

Sweat. Hypohidrosis; sweat glands small and few in number.

Skin. Generally dry; scaling with numerous bullae on the face and limbs; scanty hair follicles.

Hearing. No data.

Eyes. Photophobia; hypoplasia of the ocular puncta leading to epiphora and blepharitis.

Face. No data.

Psychomotor and growth development. EEG alterations; mild mental retardation.

Limbs. Bilateral clubfoot.

Other findings. Cleft palate.

Etiology. Autosomal incomplete dominant (AID).

Comments. The "complete" syndrome was described by Moynahan [1970] in two sibs (brother and sister) who were the offspring of first cousins. The only manifestation of the heterozygous state of this gene is an enamel defect. Were it not for the consanguineous marriage that led to the production of two homozygotes, the trait might have been considered a (provisionally) complete dominant.

Other references. None.

5. ROSSELLI-GULIENETTI'S SYNDROME (McK 22500)

Synonyms. Ectodermal dysplasia-cleft lip and palate-popliteal pterygia syndrome.

Hair. Wooly (negroid type), thin, coarse, opaque, and short.

Teeth. Hypodontia; transverse striation; irregularities of the free margins.

Nails. Subungual hyperkeratosis; sulci; transverse and longitudinal striae; thinning of the lamina; irregularities of the free margins; hallucal nails with absence of the lamina.

Sweat. In case 2, "the palmar regions showed only a mild tendency to perspiration" [Rosselli and Gulienetti, 1961; page 193]. No test was applied to evaluate hypohidrosis.

Skin. Dystrophic in the face with a tendency to desquamation (pityriasic type) and with erythematous patches; papulofollicular dermatosis in the trunk; supernumerary nipples; popliteal and perineal pterygium.

Hearing. No data.

Eyes. No data.

Face. Cleft lip; hypoplasia of the auricular lobes; flat nasal pyramid with reduced subseptum.

Psychomotor and growth development. No data.

Limbs. Aplasia or hypoplasia of thumb.

Other findings. Cleft palate; malformation of the genito-urinary system; absence or fusion of the last lumbar vertebra.

Etiology. AR.

Comments. Only cases 2 and 3 described by Rosselli and Gulienetti [1961] are considered here. Case 1 seems to represent the ectrodactyly-ectodermal dysplasia-cleft lip/palate (EEC) syndrome; no diagnosis is possible in case 4 because of incomplete documentation.

Contrary to the opinion of Pinsky [1975], cases 2 and 3 do not seem to be examples of the multiple pterygium syndrome since their several ectodermal and digital anomalies have not been observed in any case of multiple pterygium syndrome [Bixler et al, 1973]. Otherwise, the Rosselli-Gulienetti's syndrome seems to be inherited as an AR trait (the two patients are the offspring of second cousins) while the multiple pterygium syndrome is due to an AD gene.

Witkop et al [1975] suggested that cases 2 and 3 had the popliteal pterygium syndrome and an ectodermal dysplasia, ie, two different conditions at once. This seems to be improbable.

The fact that no sweat test was applied to the patients and that their possible hypohidrosis was described as mild and localized points to the possibility that this syndrome may belong to the 1-2-3 rather than the 1-2-3-4 subgroup. Further investigations are needed to clarify this point.

Other references. None.

6. DYSKERATOSIS CONGENITA (McK 22423 and 30500)

Synonyms. Dyskeratosis congenita with pigmentation, dystrophia unguium, and leukoplakia oris; Zinsser-Engman-Cole's syndrome; Cole-Rauschkolb-Toomey's syndrome; pigmentatio parvo-reticularis cum leukoplakia et dystrophia unguium.

Hair. Hypotrichosis; loss of cilia due to blepharitis and ectropion; absence of eyebrows and lashes; premature canities.

Teeth. Poorly aligned; early carious degeneration.

Nails. Dystrophy with late onset (childhood and puberty); paronychia occasionally leading to anonychia; hypoplasia.

Sweat. Palmoplantar hyperhidrosis; generalized hypohidrosis elsewhere.

Skin. Hyper- and hypomelanosis; telangiectatic erythema; acrocyanosis; bullae; ulcers; dry desquamation; atrophy; hyperkeratotic plaques (palmoplantar and over joints); premalignant lesions.

Hearing. No data.

Eyes. Blepharitis; ectropion of the lower lids; obliteration of the puncta lacrimalia; bullous conjunctivitis.

Face. Sharp-featured.

Psychomotor and growth development. Occasional mental and growth retardation.

Limbs. Normal.

Other findings. Fanconi-like pancytopenia; premalignant leukoplakia on lips, mouth, anus, urethra, and conjunctiva; frail skeletal structure; genital anomalies; esophageal dysfunction and/or diverticulum; atrophic lingual papillae; gingivitis.

Etiology. Heterogeneity. XR, AR, and AD.

Comments. Bopp and Bernardi [1974] described three affected men and called attention to the frequent "association" of this condition with Fanconi's anemia. De Boeck et al [1981] described a case and called attention to the fact that, although dyskeratosis congenita is a rare condition, it should be considered as a diagnostic possibility for any patient with aplastic anemia or thrombocytopenia. This association has been mentioned frequently [McKusick, 1983].

The claimed occurrence of dyskeratosis congenita and Fanconi's anemia in the same patient cannot be attributed to concurrence (or syntropy). What seems plausible is a clinical overlap of the two conditions: over 50% of patients with dyskeratosis congenita also develop a Fanconi-like hematopoietic picture [De Boeck et al, 1981]. The name Zinsser-Fanconi's syndrome has been erroneously applied to these cases. It seems absolutely clear that Fanconi's syndrome (AR) and dyskeratosis congenita (AR, AD, XR) are two clinically well-delineated conditions in spite of the fact that there are patients with the latter condition together with a hematologic picture similar to that of Fanconi's syndrome. The presence of mucocutaneous signs and the hematologic picture "is typical for dyskeratosis congenita, but usually the cutaneous manifestations appear first" [De Boeck et al, 1981].

Etiology of dyskeratosis congenita is commonly mentioned as heterogeneous, but only XR and AR inheritance are generally mentioned. However, a few papers described kindreds presenting a clear AD inheritance. Degos et al [1969], for example, described a proposita, her sister, her father, her paternal uncle and her paternal grandmother. Scoggins et al [1971] described two men and four women in three generations: the propositus, his three sisters, his father and his daughter [cf personal communication to Sirinavin and Trowbridge, 1975].

The paper by Nazzaro et al [1972] (with the description of two brothers and five other relatives, two of whom were women) deserves a special comment. The relationship of the five relatives is not presented, the condition (with variable degrees of severity) is only described in the two brothers, and the conclusion is reached that the pedigree suggests an X-linked incomplete dominance. With such an incomplete account, it is possible that this French kindred also represents an example of AD inheritance.

Other references. Jansen [1951], Garb [1958], Sorrow et al [1963], Bryan and Nixon [1965], Steier et al [1972], Witkop et al [1975].

7. PACHYONYCHIA CONGENITA (McK 16720)

Synonyms. Jadassohn-Lewandowsky's syndrome; pachyonychia ichthyosiforme; polykeratosis congenita; pachyonychia congenita types I, II, and III; pachyonychia neonatorum.

Hair. Occasional alopecia and dystrophic changes; dry.

Teeth. Natal and carious; early (under 30) loss of secondary teeth due to caries; occasional early eruption of deciduous teeth.

Nails. Greatly thickened and angling upward on all digits; the free edges are distally raised, narrowed and pointed; paronychia.

Sweat. Palmoplantar hyperhidrosis.

Skin. Patchy to complete hyperkeratosis of palms and soles; easily blistering foot callosities; epidermoid cysts; verrucous lesions on the knees, elbows, buttocks, ankles, and popliteal regions.

Hearing. Normal.

Eyes. Corneal thickening; cataract.

Face. Normal.

Psychomotor and growth development. Normal.

Limbs. Normal.

Other findings. Oral and mucosal keratosis; osteomata; intestinal diverticuli.

Etiology. AD.

Comments. Three clinical forms have been described: type I, symmetrical keratosis of the hands and feet with follicular keratosis of the trunk; type II, type I plus oral mucosal lesions resembling leukokeratosis or leuko-

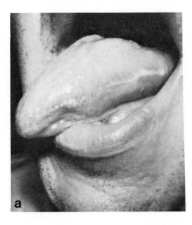 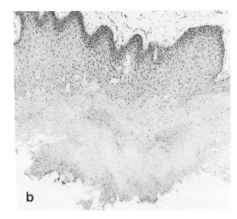

Fig. 3-3. Pachyonychia congenita. a. Hyperkeratotic lesions at the comissures of the lip and lateral border of the tongue. b. Photomicrograph of a lingual biopsy of the same patient showing horny hyperkeratosis. Courtesy of Dr. Carl J. Witkop, Jr., Minneapolis, MN.

plakia and scalloped tongue (this is the most frequent, the so-called Riehl type); type III, type I with corneal changes. The three forms may be seen in the same kindred.

Stieglitz and Centerwall [1983] reviewed the literature and added an analysis of 17 patients, with an overall total of 167. They reported some unusual findings (respiratory and dental involvement) and the possibility of distorted segregation ratios. Since this condition generally does not exhibit manifestations at birth (with the exception of occasional natal teeth), these authors say that the term pachyonychia congenita is misleading; they suggest pachyonychia neonatorum for it.

According to Greene [1962], dyskeratosis congenita and pachyonychia congenita are so similar that they may represent different forms of the same condition. However, this seems to be an isolated opinion. These conditions are generally accepted as two different syndromes.

Other references. Wright and Guequierre [1947], Jackson and Lawler [1951], Witkop and Gorlin [1961], Joseph [1964], David et al [1977], Schönfeld [1980].

8. RAPP-HODGKIN'S SYNDROME (McK 12940)

Synonyms. Hypohidrotic ectodermal dysplasia with cleft lip and cleft palate; anhidrotic ectodermal dysplasia with cleft lip and cleft palate.

Hair. Coarse and stiff on scalp; pili torti; absence or scarcity on scalp and body; sparse eyebrows and lashes.

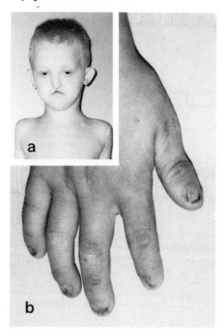

Fig. 3-4.   Rapp-Hodgkin's syndrome. a. Scalp hypotrichosis, scanty eyebrows, repaired cleft lip, and abnormal and prominent auricles. b. Onychodystrophy. Courtesy of Dr. Robert L. Summitt, Memphis, TN.

Teeth. Conically shaped; short, square incisors and canines; hypoplastic enamel; extensive caries; hypodontia.

Nails. Small, narrow, and dysplastic.

Sweat. Hypohidrosis; lower number of sweat glands.

Skin. Dry and coarse; thickened over the extensor surface of the elbows and knees; hypoplastic dermatoglyphics.

Hearing. Conductive loss (secondary to otitis media) in three patients.

Eyes. Chronic epiphora; corneal opacities; photophobia; atresia of puncta; ectropion; lacrimal papillae.

Face. Cleft lip; hypoplastic maxilla; mild frontal prominence; microstomia; mildly depressed nasal bridge; prominent and malformed auricles.

Psychomotor and growth development. Short stature.

Limbs. Occasional syndactyly.

Other findings. Cleft palate; chronic rhinitis; nasal speech; hypospadias.

Etiology. AD? XD? See *Comments.*

Comments. Rapp and Hodgkin [1968] described a mother, her son, and daughter. Summitt and Hiatt [1971] described a boy (sporadic case). The inheritance is probably AD, although XD is not excluded. Wannarachue et al [1972] reexamined Rapp and Hodkin's patients. The condition observed in the girl reported by Beckerman [1973] is similar to this syndrome.

Other references. Stasiowska et al [1981], Silengo et al [1982].

9. ECTRODACTYLY-ECTODERMAL DYSPLASIA-CLEFT LIP/ PALATE (EEC) SYNDROME (McK 12990)

Synonym. Ectrodactyly-ectodermal dysplasia-clefting syndrome.

Hair. Hypotrichosis of scalp and body; fair and dry; scanty or absent eyebrows and lashes.

Teeth. Anodontia; hypodontia; microdontia; enamel hypoplasia; poorly formed; increased caries; peg-shaped incisors.

Nails. Dysplastic, thin, brittle, and striated; pitted and terminated irregularly.

Sweat. Occasional hypohidrosis without hyperthermia.

Skin. Dry, translucent, dystrophic; palmoplantar hyperkeratosis; eczematous patches; pigmented nevi.

Hearing. Conductive loss.

Eyes. Tear duct anomaly or malfunction; speckled iris; photophobia; strabismus; blepharitis; clouding of the cornea; congenital adhesions between the eyelids.

Face. Cleft lip; broad nose; defective auricles (small and low-set; lack of the usual amount of cartilage; posteriorly rotated); pointed chin; malar hypoplasia.

Psychomotor and growth development. Mental retardation.

Limbs. Ectrodactyly (split hands/feet); syndactyly; clinodactyly.

Other findings. Cleft palate; renal abnormalities; rhinitis; respiratory infections; genital anomalies.

Etiology. AD.

Comments. According to Preus and Fraser [1973], tear duct anomaly or malfunction is more frequent than cleft lip, and cleft lip-palate may not be an essential part of this syndrome.

EEC syndrome has been identified with the odontotrichomelic syndrome (OS) by some investigators [Brill et al, 1972; Rosenmann et al, 1976; Gemme et al, 1976; Schnitzler et al, 1978]. However, others [Freire-Maia, 1971, 1977a; Preus and Fraser, 1973; Witkop et al, 1975; Pinsky, 1975, 1977; Rapone-Gaidzinski, 1978; Cohen, 1978, 1979] did not agree

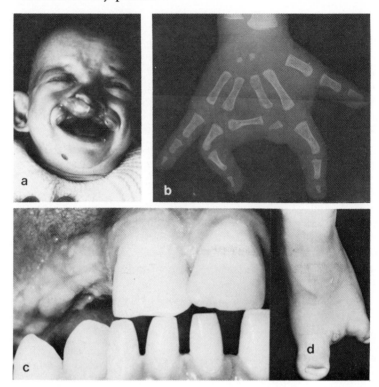

Fig. 3-5.   Ectrodactyly-ectodermal dysplasia-cleft lip/palate syndrome. a. Cleft lip and palate; abnormal auricle. b. Ectrodactyly. c. Hypodontia, peg-shaped and widely spaced teeth. d. Ectrodactyly (split feet). a,b, courtesy of Dr. Roland Walbaum, Lille, France; c,d, courtesy of Dr. Geoffrey C. Robinson, Vancouver, BC, Canada.

with this identification. In spite of some similarities (as between other ectodermal dysplasias), the differences suggest that it is more reasonable to treat them as separate entities [Pinheiro and Freire-Maia, 1980]. The main differences between the two syndromes are related to limb anomalies.

Brill et al [1972] and Rosenmann et al [1976] have suggested that genetic heterogeneity may exist in the EEC syndrome. Two AD forms have been postulated: One with cleft lip with or without cleft palate and another with cleft palate alone. Bowen and Armstrong [1976, 1979] reported three sibs with a condition with overlapping manifestations of EEC, OS, and other conditions; they considered it as a separate entity and suspected that some patients reported to have EEC syndrome may have the same condition they described. Witkop et al [1975] considered Bowen's syndrome to be

EEC syndrome. Pinsky [1977] identified Bowen's syndrome as AEC syndrome (see below).

Other references. Rüdiger et al [1970], Bixler et al [1972], Fried [1972], Swallow et al [1973], Robinson et al [1973].

10. ANKYLOBLEPHARON-ECTODERMAL DEFECTS-CLEFT LIP AND PALATE (AEC) SYNDROME (McK: not listed)

Synonym. Ankyloblepharon and ectodermal dysplasia.

Hair. Hypotrichosis (from "dystrophic" and "sparse and brittle" to almost totally absent scalp, axillary and pubic hair); absent or scanty eyebrows and lashes.

Teeth. Poorly formed and pointed; widely spaced; carious and discolored; severe hypodontia; delayed eruption.

Nails. Severe dystrophy (the abnormalities vary in extent from only terminal dystrophy to total absence).

Sweat. Hypohidrosis without hyperthermia; lower number of sweat pores.

Skin. Dry and smooth; palmoplantar hyperkeratosis with obliteration of dermatoglyphic patterns; occasionally reticulate hyperpigmentation; supernumerary nipples.

Hearing. Neural loss in one patient.

Eyes. Lacrimal duct atresia; photophobia.

Face. Ankyloblepharon filiforme adnatum with partial fusion of eyelids at birth; broadened nasal bridge; hypoplastic maxilla; auricular abnormalities; cleft lip.

Psychomotor and growth development. Normal.

Limbs. Syndactyly.

Other findings. Cleft palate. One patient had "a number of filamentous bands joining the lateral walls of the posterior aspect of the vagina" and "a posterior anal fissure" (page 283).

Etiology. AD.

Comments. Seven patients (five females and two males) from four families were described by Hay and Wells [1976]: three cases of transmission from parent to offspring (six patients) and a sporadic case. See *Comments* on EEC syndrome.

Other reference. Pinsky [1977].

11. ZANIER-ROUBICEK'S SYNDROME (McK: not listed)

Synonyms. None.

Hair. Hypotrichosis (more striking in men); normal eyebrows and lashes.

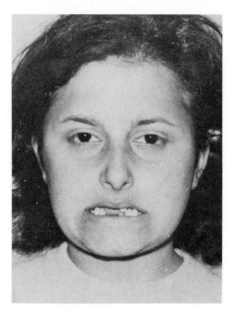

Fig. 3-6. Zanier-Roubicek's syndrome. Hypodontia and coniform teeth, thin nose, hypoplasia of alae nasi, and apparent hypotelorism. Courtesy of Dr. Martin Roubicek, Mar del Prata, Argentina.

Teeth. Hypodontia; conical teeth; early loss of deciduous teeth; enamel with yellow transverse streaks.

Nails. Occasionally brittle.

Sweat. Hypohidrosis often with severe hyperthermia in infancy; lower number of sweat glands; normal sweating on palms and soles.

Skin. Smooth and dry.

Hearing. Normal.

Eyes. Reduced lacrimation.

Face. See *Comments.*

Psychomotor and growth development. Normal or slightly reduced stature. One patient had a convulsive disorder with abnormal EEG attributed to hyperthermia in infancy; she also had a transient paresis, which improved spontaneously.

Limbs. Normal.

Other findings. Hypoplasia of the mammary glands.

Etiology. AD.

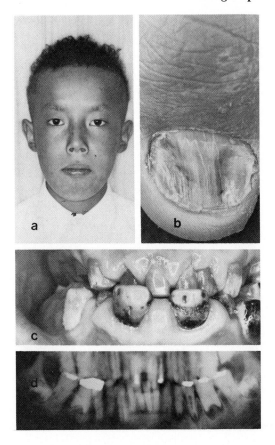

Fig. 3-7. Trichoonychodental syndrome. a. Hypotrichosis, curly and fine scalp hair, scanty eyebrows and lashes. b. Hallux nail showing discrete central elevation, longitudinal striations, and splitting. c. Mottled, striated hypoplastic-hypomature enamel with swelling of the interdental papillae. d. Taurodontism, radiolucent areas around the apices, arrow nock appearance of the single-rooted teeth, obliteration of pulp chambers and lack of contrast between enamel and dentin. Courtesy of Dr. Hiroaki Koshiba, Tokyo, Japan.

Comments. Zanier and Roubicek [1976] described one family with 21 affected persons (11 men). The above clinical findings are based only on the two men and two women who were examined.

Figure 3-6 shows a woman with narrow nose, hypoplasia of alae nasi, and apparent hypotelorism. Asked to provide information on these traits, the authors informed us that hypoplasia of alae nasi and narrow nose may be accepted as occasional and mild signs of this syndrome; however, hypotelorism is only apparent [Zanier and Roubicek, 1982, pers. comm.].

Other reference. Roubicek [1979, pers. comm.].

12. TRICHOONYCHODENTAL (TOD) DYSPLASIA (McK: not listed)

Synonyms. Familial congenital ectodermal dysplasia; odontogenesis imperfecta.

Hair. Scanty, fine, and curled; sparse eyebrows and lashes.

Teeth. Taurodontic molars; hypoplastic-hypomature enamel; dysplasia of dentin; hypodontia; persistence of deciduous teeth; widely spaced.

Nails. Thin, with a central elevation and longitudinal striations and cracks. Both fingernails and toenails are affected but the toenails are more severely involved.

Sweat. Hypohidrosis with hyperthermia; slightly decreased number of dermal pores.

Skin. Fine texture.

Hearing. Normal.

Eyes. Normal.

Face. Normal.

Psychomotor and growth development. Normal.

Limbs. Normal.

Other findings. No data.

Etiology. AD.

Comments. Koshiba et al [1978] described a kindred of Japanese ancestry with six affected members (four men) over three generations.

The first of the above synonyms is too vague to be of any value; the second does not describe the condition adequately.

Other references. None.

13. JORGENSON'S SYNDROME (McK: not listed; see *Comments*)

Synonyms. Ectodermal dysplasia with hypotrichosis, hypohidrosis, defective teeth and unusual dermatoglyphics.

Hair. Coarse scalp hair that grows slowly and sheds rapidly toward the end of the second decade of life; sparse eyebrows and lashes.

Teeth. Deciduous teeth with brown spots and susceptible to decay; congenital absence of lower permanent lateral incisors.

Nails. Short and thick, with longitudinal ridges.

Sweat. Hypohidrosis (the patients sweat evenly over their entire body, although the amount of sweat is less than normal).

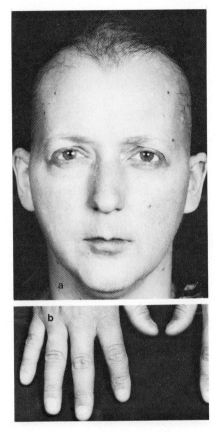

Fig. 3-8. Jorgenson's syndrome. a. Woman with hypotrichosis of scalp, sparse eyebrows (especially on the sides), sparse lashes, pinched nares, thin upper lip, long philtrum, and somewhat masculine features. b. Dry skin and dystrophic nails. Courtesy of Dr. Ronald J. Jorgenson, Charleston, SC.

Skin. Generally dry; smooth and thick over the distal phalangeal joints; fine dermal ridges on the hands and feet; single palmar flexion creases; keratosis pilaris on both knees.

Hearing. Decreased acuity on the left side (?).

Eyes. The conjunctivae are dry and conjunctivitis is frequent.

Face. Strikingly similar; characterized by thin alae nasi, long philtrum, and thin upper lip; narrow palpebral fissures.

Psychomotor and growth development. Normal.

Limbs. Normal.

Other findings. Hyperextensible joints; posterior fusion of labia majora; scarce vaginal secretion.

Etiology. AD? XD? See *Comments.*

Comments. Jorgenson [1974] described three women belonging to three generations of a family. The mode of transmission is probably AD, although an XD hypothesis cannot be excluded. Contrary to the opinion of Jorgenson [1974] and McKusick [1983; McK 12920], the family described by Basan [1965] possibly has another condition.

Other references. None.

14. CAREY'S SYNDROME (McK: not listed)

Synonyms. None.

Hair. Thin, hypopigmented, and very sparse; poor growth.

Teeth. Abnormally formed with discoloration; microdontia; hypodontia (at the age of 25, all teeth were either capped or replaced with dentures).

Nails. Dystrophic from early childhood.

Sweat. Decreased number of sweat pore openings. The patient thought that she sweated much less than normal, but she was able to endure extremely hot weather comfortably.

Skin. Aplasia cutis congenitalike scalp defects.

Hearing. Moderate conductive hearing loss.

Eyes. Absence of tear ducts; displacement of the inner canthi.

Face. U-shaped mouth; flat nasal bridge; maxillary hypoplasia.

Psychomotor and growth development. Normal.

Limbs. Incomplete 2-3 toe syndactyly. See *Comments.*

Other findings. Cleft palate.

Etiology. Unknown.

Comments. Carey [1980, pers. comm.] described one woman from normal nonconsanguineous parents. Incomplete 2-3 toe syndactyly is a common minor anomaly; it may not be a sign of the syndrome.

Other references. None.

15. CAMARENA SYNDROME (McK: not listed)

Synonym. Partial anhidrotic ectodermal dysplasia with shortness of the left leg.

Hair. Thin and brittle; hypertrichosis; absent eyebrows; scanty eyelashes.

Teeth. Anodontia.

Nails. Dysplastic.

Sweat. Absence of sweat glands in the scalp (biopsy); anhidrosis on the face and scalp; euhidrosis on the rest of the body.

Skin. Thin and smooth; palmoplantar erythema; nevus vascularis on the right lid and above the nose.

Hearing. Normal.

Eyes. Normal.

Face. Mild "cara de vieja" (old woman's face); hypertelorism; abnormal auricles; micrognathia; microstomia.

Psychomotor and growth development. Normal.

Limbs. Bilateral clinodactyly of the fifth fingers; left femur and tibia about 0.8 cm and 0.7 cm shorter than the right ones, respectively.

Other findings. Highly arched palate.

Etiology. AD? XD?

Comments. Felix Rodriguez et al [1980] described four persons with the "complete form" (the proband was described above) and eight patients with the "incomplete form" (thin and brittle hair; malocclusion with hypodontia) in three generations of one family.

The shortness of the left leg may be coincidental since it was found only in the proband.

The name of the syndrome is that of the Spanish town where the patients were born.

Other references. None.

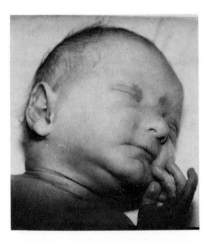

Fig. 3-9. Camarena syndrome. Scalp hypotrichosis, absent eyebrows, scanty lashes, abnormal auricles and capillary hemangioma on the right lid and above the nose. Note palmar erythema. Courtesy of Dr. Valentin Felix Rodriguez, Toledo, Spain.

16. ICHTHYOSIFORM ERYTHRODERMA-DEAFNESS-KERATITIS (McK 24215)

Synonyms. Keratitis-ichthyosis-deafness (KID) syndrome; congenital ichthyosiform syndrome with deafness and keratitis; ichthyosiform erythroderma and congenital neurosensory deafness; Wilson-Grayson-Pieroni's syndrome; icthyosiform erythroderma-corneal involvement-deafness; spiny hyperkeratosis-alopecia-deafness; Senter's syndrome.

Hair. Hair loss varies from alopecia to fine, thin scalp hair; scanty or absent eyebrows and lashes; occasional trichorrhexis nodosa in some scalp hairs; alopecia may occur, but only during the first or second year of life.

Teeth. Delayed eruption of deciduous teeth; brittleness; tendency to develop caries; unspecified defects.

Nails. Absent at birth; delayed development; leukonychia and thickening (most marked in the fingernails); destructive dystrophy.

Sweat. Hypohidrosis (with hyperthermia) due to the low number of sweat glands.

Skin. Ichthyosiform erythroderma with sebaceous dysfunction; furrowing around mouth and chin; erythematous hyperkeratotic plaques on elbows, knees, and the dorsa of hands and feet; marked thickening (leatherlike consistency) of palms and soles.

Hearing. Congenital sensorineural deafness.

Eyes. Vascularization of the corneae with pannus formation resulting in loss of vision; keratitis; occasionally decreased tear production; photophobia.

Face. Normal.

Psychomotor and growth development. Occasional growth deficiency.

Limbs. Bilateral flexion contractures at knees and elbows with tight heel cords.

Other findings. No data.

Etiology. AR.

Comments. According to Cram et al [1979], at least 16 cases (eight men) have been reported. Desmons et al [1975] reported three affected siblings whose parents were first cousins.

Other references. Morris et al [1969], Myers et al [1971], Wilson et al [1973], Rycroft et al [1976], Senter et al [1978], Skinner et al [1981].

17. ANONYCHIA WITH BIZARRE FLEXURAL PIGMENTATION (McK 10675)

Synonym. Anonychia with flexural pigmentation.

Hair. Slow-growing and coarse scalp hair, thinning early in adult life.

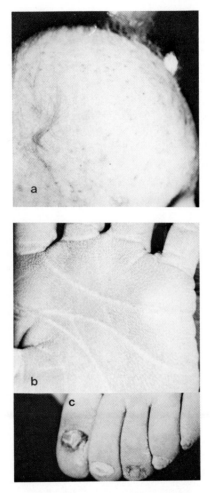

Fig. 3-10.   Ichthyosiform erythroderma-deafness-keratitis. a. Severe hipotrichosis, come-dones, and acne. b. Palmar hyperkeratosis. c. Severe onychodystrophy. Courtesy of Dr. Merrill Grayson, Indianapolis, IN.

Teeth. Highly carious. One patient had all teeth extracted by the time he was 18 years old.

Nails. Generally absent on the fingers and toes; in a few instances, rudimentary.

Sweat. Mild hypohidrosis without hyperthermia.

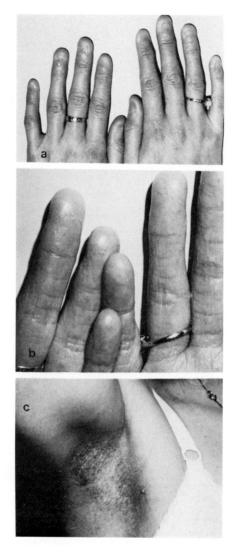

Fig. 3-11.   Anonychia with bizarre flexural pigmentation. a. Anonychia and dry skin. b. Keratosis; excess of flexion creases. c. Pigmentation at the axilla. Courtesy of Dr. Julian Verbov, Liverpool, England.

Skin. Hypo- and hyperpigmentation, particularly in the groins, axillae, and breasts; distortion of epidermal ridges on palms and soles; mild palmoplantar hyperkeratosis; increased palmar markings; distorted finger-tip patterns; small macular telangiectases in a few regions.

Hearing. No data.

Eyes. No data.

Face. "Hooked" nose.

Psychomotor and growth development. No data.

Limbs. Normal.

Other findings. No data.

Etiology. AD.

Comments. Verbov [1975] described four persons (two men) over two generations of one family.

Other references. None.

18. HYPOHIDROTIC ECTODERMAL DYSPLASIA WITH PAPILLOMAS AND ACANTHOSIS NIGRICANS (McK: not listed)

Synonyms. None.

Hair. Generalized hypotrichosis; scalp hair is dry, fine, lusterless, woolly (in two patients), and slow-growing (has not grown more than 2–3 cm); similar involvement in the axillary and pubic areas; scanty and slow-growing moustache and beard (affected men shave once a month); scanty eyebrows and lashes.

Teeth. Hypoplastic and carious; anodontia at the age of 30 in one patient.

Nails. Short and dystrophic.

Sweat. Hypohidrosis; small number of sweat glands (detected by biopsy).

Skin. Dry; palmoplantar hyperkeratosis; hyperpigmented and hyperkeratotic skin with wrinkles, papillomas, and a symptomatic acanthosis nigricans in the neck, axillae, and genito-femoral regions; increased cornification and presence of follicular plugs on the "normal" skin; unusual wrinkles around lips.

Hearing. No data.

Eyes. No data.

Face. Normal.

Psychomotor and growth development. Mild mental retardation (one patient).

Limbs. No data.

Other findings. Plicate tongue with papillomatosis.

Etiology. AR?

Comments. Lelis [1978] described four persons (two men) from normal parents. The other members of their families were reported to be normal but no information on inbreeding was provided.

Other references. None.

19. ODONTOONYCHOHYPOHIDROTIC DYSPLASIA WITH MIDLINE SCALP DEFECT (McK: not listed)

Synonyms. None.

Hair. Secondary alopetic area.

Teeth. Delayed eruption; diastemata; minor shape alterations.

Nails. Dystrophic fingernails.

Sweat. Hypohidrosis as determined by pilocarpine iontophoresis.

Skin. Midline scalp defect (aplasia cutis verticis); hypoplastic areolae and nipples.

Hearing. Normal.

Eyes. Normal.

Face. Normal.

Psychomotor and growth development. Normal.

Limbs. Normal.

Other findings. Breast hypoplasia (inability to lactate); hypertension of undetermined pathogenesis; renal abscess in the mother; adrenal cortical cyst in the son.

Etiology. AD.

Comments. Tuffli and Laxova [1983] described a woman and one of her two sons; no definite indication of involvement observed in five brothers and three sisters. Whether adrenal cortical cyst, renal abscess, and hypertension are components of the condition or mere coincidental findings remains to be determined.

Other reference. Tuffli [1981, pers. comm.].

20. ODONTOONYCHODERMAL DYSPLASIA (McK: not listed)

Synonyms. None.

Hair. Normal to sparse on scalp; dry; normal to sparse eyebrows; normal lashes; normal axillary and pubic hair.

Teeth. Hypodontia; hypoplastic. Coniform teeth, sometimes with pointed edges and peculiar, bent form.

Nails. Short, thickened, and brownish toenails; normal to mildly dystrophic fingernails. In one patient, no nail changes were observed.

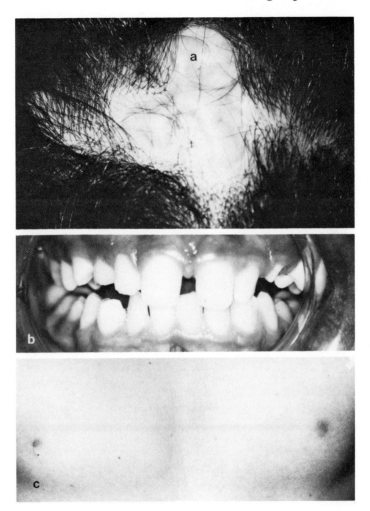

Fig. 3-12.   Odontoonychohypohidrotic dysplasia with midline scalp defect. a. Large aplasia cutis verticis defect. b. Dental defects. c. Hypoplastic nipples and areolae. Courtesy of
Dr. Gordon A. Tuffli, Madison, WI.

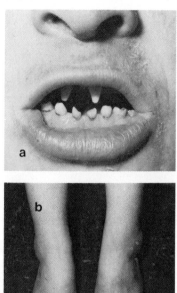

Fig. 3-13.   Odontoonychodermal dysplasia. a. Skin and teeth involvement. b. Onychodystrophy. Courtesy of Dr. Vazken M. Der Kaloustian, Beirut, Lebanon.

Sweat. Palmoplantar hyperhidrosis.

Skin. Palmoplantar erythema, hyperkeratosis, and fissuring. Erythema, atrophy, telangiectasia, and scaliness on the face. Scaly dry skin. Normal nipples and areolae. Normal hair follicles and eccrine glands. Nevus flammeus of right leg in one patient. Basophilic dermis degeneration compatible with solar elastosis. Decreased number and size of sebaceous glands.

Hearing. Normal.

Eyes. Normal.

Face. Normal.

Psychomotor and growth development. Normal.

Limbs. Normal.

Other findings. None.

Etiology. AR.

Comments. Fadhil et al [1983] described three inbred (F = 1/16) sibships from two apparently different Lebanese families with seven affected persons (five men) out of a total of 24.

Other references. None.

21. TRICHOODONTOONYCHO-HYPOHIDROTIC DYSPLASIA WITH CATARACT (McK: not listed; see *Comments*)

Synonyms. None.

Hair. Almost complete alopecia. No hair is visible over the body, but the scalp has occasional long, fine lanugolike hairs (both patients).

Teeth. Widely spaced, hypoplastic, and conical small teeth, with the enamel pitted in places (one patient).

Nails. Hypoplastic (both patients).

Sweat. Hypohidrosis with hyperthermia. Sweat glands are poorly and irregularly developed (both patients).

Skin. Mottled with dull, red, confluent, slightly depressed and mildly indurated zones that blanch very slightly with pressure. Between these darker regions, there are small, whitish, and ill-defined 3- to 10-mm areas. The changes are most accentuated over the limbs, head, and back. Irregular hypoplasia of sebaceous glands (similar in both patients).

Hearing. No data.

Eyes. Bilateral cataract (of the lamellar type) verified in infancy (both patients); moderate enophthalmos (one patient); bilateral esotropia (both patients).

Face. Depressed nasal bridge; frontal bossing; inner epicanthic folds are thickened and webbed over the medial part of both eyes (both patients).

Psychomotor and growth development. Both patients retarded; hypoactive deep reflexes (one patient).

Limbs. Normal.

Other findings. Highly arched palate (one patient); atrophic nasal mucosae (both patients).

Etiology. AR? See *Comments*.

Comments. Cole et al [1945] described two sisters from normal nonconsanguineous parents. The sibship also includes two normal girls.

On the basis of the fact that the father has slightly conical incisors, which are more widely spaced than usual, Cole et al [1945] admit an incomplete dominant pattern of inheritance. Since the rest of the family, including their paternal grandparents, is normal, we prefer to assume an AR? cause.

The paper by Cole et al [1945] is referred to by McKusick [1983] in the bibliography of the entry on Rothmund-Thomson's syndrome (McK 26840).

Other references. None.

22. PAPILLON-LEFÈVRE'S SYNDROME (McK 24500)

Synonyms. Hyperkeratosis palmoplantaris with premature periodontoclasia; keratosis palmoplantaris with periodontopathia; parodontopathia acroectodermalis; palmar-plantar hyperkeratosis and periodontal destruction syndrome.

Hair. Occasionally thin and "loose."

Teeth. Periodontal degeneration (periodontoclasia) of both primary and secondary dentition with consequent shedding of all teeth (generally lost around 12–15 years of age); occasional enamel hypoplasia.

Nails. Occasionally dystrophic (spoon-shaped and striated; onychogryposis).

Sweat. Palmoplantar hyperhidrosis; occasional generalized hypohidrosis.

Skin. Hyperkeratosis of the palmar and plantar surfaces with a tendency toward fissuring and cracking; dry and dirty-appearing on the dorsal surface of the arms and the ventral surface of the legs; occasional eczema and erythema of the face as well as of the sacral and gluteal regions.

Hearing. Normal. Occasional congenital deafness (?).

Eyes. Occasional microphthalmia.

Face. Normal. Occasional frontal bossing (?).

Psychomotor and growth development. Occasional retardation.

Limbs. Occasional arachnodactyly and other bone abnormalities.

Other findings. Severe gingivostomatitis; occasional intracranial calcifications; abnormal liver function; renal abnormalities; generalized osteoporosis.

Etiology. AR.

Comments. Gorlin et al [1964] published an analysis of 46 well-documented cases and suggested that calcification of the dura mater is also a component of the syndrome. Laynes-de-Andrade [1974] described four cases and reviewed 84 of 113 reported in the literature.

According to McKusick [1979, 1983], the two daughters from first-cousin parents described by Schöpf et al [1971] may have this syndrome (see "Cystic eyelids, palmoplantar keratosis, hypodontia and hypotrichosis").

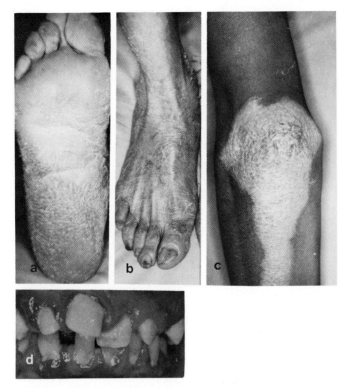

Fig. 3-14. Papillon-Lefèvre's syndrome. a. Severe plantar hyperkeratosis; note cratered aspect. b,c. Transgressive keratotic lesions. d. Enamel hypoplasia of the upper incisors, malposition, peg-shaped, and widely spaced teeth. Courtesy of Dr. Fernando Laynes-de-Andrade, Curitiba, PR, Brazil.

Other references. Ingle [1959], Ziprokowski et al [1963], Picarelli [1968], Carvel [1969], Piguet et al [1969], Giansanti et al [1973], Munford [1976], Haneke [1979].

23. HYPOMELANOSIS OF ITO (McK 14615)

Synonym. Incontinentia pigmenti achromians.
Hair. Diffuse alopecia; facial hypertrichosis.
Teeth. Dysplastic; pointed extra cusps at the palatal face of maxillary incisors; hypodontia.
Nails. Transversal ridging.
Sweat. Hypohidrosis as verified by pilocarpine test; relative hyperhidrosis at the hypopigmented areas.

Skin. Asymmetric and bizarre hypopigmented areas in different forms (linear, guttate, whorled) in the whole body with exception of the scalp, palms, soles, and mucous membranes. The skin involvement may appear at or shortly after birth, or even during childhood with a tendency to progression to uninvolved regions. The bandlike hypopigmented areas generally parallels without crossing the midline and have a tendency to return to normal color.

Hearing. Conductive loss.

Eyes. Myopia, strabismus, retinal pigmentary abnormalities, choroidal atrophy, opaque cornea, pupillary dislocation, tessellated fundus, microphthalmia, iridial heterochromia, astigmatism, optic atrophy, etc.

Face. Epicanthal folds, palpebral ptosis, thick lips, hemiatrophy, hypertelorism, malformed auricles, cleft lip, saddle nose, etc.

Psychomotor and growth development. Motor and mental retardation, seizures, EEG abnormalities, language retardation, short stature, etc.

Limbs. Atrophy of one finger, clinodactyly, polydactyly, etc.

Other findings. Cleft palate, asymmetric head, scoliosis, diminution of capillary resistance, etc.

Etiology. AD. See *Comments.*

Comments. Although this is not a purely cutaneous condition, the large number of other involvements (individually occasional) should not obscure the fact that the skin component is the only single criterion for a reliable diagnosis [Jelinek et al, 1973]. On the basis of a review of 73 cases, Takematsu et al [1983] verified that 26% of them had only skin involvement and that the most common of the other involvements (motor and mental retardation) were present in only about one-fourth of the cases. Most of the "associated findings" were present in less than 3% of the sample.

Although this is an ectodermal dysplasia of Group A according to our definition (Chapter 2), the frequency of its associated ectodermal anomalies is lower than that of the classical incontinentia pigmenti [Rubin, 1972].

On the basis of the considerations by Jelinek et al [1973] we prefer to call this condition hypomelanosis of Ito instead of incontinentia pigmenti achromians; this last designation seems to mean "incontinentia pigmenti in process of decoloring" (page 601), which would presuppose the preexisting presence of incontinentia pigmenti.

For differential diagnosis, it is important to remember that this condition "appears like a negative image of incontinentia pigmenti" (page 597), and that there is neither verrucous changes nor inflammation preceding the depigmentation [Jelinek et al, 1973].

The cause is generally assumed to be an AD gene. However, it is important to note that both among Japanese and non-Japanese patients, the sex ratio (respectively 23:9 and 28:10; total 51:19; in favor of women) departs widely from the expected 1:1 ratio (the corresponding χ^2 values are 6.12, 8.52, and 14.62, respectively). This discrepancy points either to heterogeneity or to a mechanism distorting the expected Mendelian ratio.

Other references. None.

4. Subgroup 1-2-3

1. ROTHMUND-THOMSON'S SYNDROME (McK 26840)

Synonyms. Poikiloderma congenita; poikiloderma congenitale; Rothmund's syndrome; Rothmund's dystrophy; Thomson's syndrome; telangiectasis-pigmentation-cataract syndrome; Block-Stauffer's dyshormonal dermatosis; congenital cutaneous dystrophy; congenital poikiloderma-juvenile cataract syndrome: poikiloderma atrophicans and cataract syndrome; Rothmund-Petges-Cléjat's syndrome.

Hair. Hypotrichosis of scalp and body; occasional alopecia; eyebrows and lashes usually fall out during the first year of life and remain sparse or absent thereafter.

Teeth. Hypodontia; microdontia; supernumerary teeth; pronounced caries; delayed eruption. See *Comments.*

Nails. Frequently dystrophic (rough, ridged, heaped-up, or atrophic).

Sweat. Normal. See *Comments.*

Skin. Lesions of poikilodermal aspects including atrophy, irregular pigmentation, and telangiectasias beginning during the first 3 to 6 months; hyperkeratosis of palms and soles; sensitivity to sunlight. At first the rash consists of tense, red, elevated, and edematous patches appearing symmetrically on the cheeks, hands, forearms, and buttocks, and subsequently on the trunk and lower limbs. After a few years, the active phase persists and a dry, scaling, and atrophic skin develops with areas of hyperpigmentation, hypopigmentation, and telangiectasia.

Hearing. No data.

Eyes. Cataract, usually bilateral (onset generally between 3 and 6 years); occasional degenerative lesions of the cornea.

Face. Occasional saddle nose, frontal bossing, and microcephaly.

Psychomotor and growth development. Short stature; occasional mental retardation.

Limbs. Small hands and feet, short terminal phalanges, syndactyly, absence of metacarpals, rudimentary ulna and radius, pelvic anomalies,

etc, have been described by different authors. (For other skeletal anomalies, see *Other findings*).

Other findings. Hypogonadism; cryptorchidism; skull abnormalities; scoliosis. Women often have amenorrhea and are often sterile.

Etiology. AR. See *Comments*.

Comments. According to Kirkham and Werner [1975], "the patients often
show marked temperature instability because of inability to sweat" (page
11). Since no other author corroborates this information, we prefer to
classify this syndrome as belonging to the 1-2-3 subgroup and not to the 1-
2-3-4 subgroup.

This syndrome is sometimes considered to represent two similar conditions (heterogeneity). However, the general tendency is to accept it as only
one syndrome.

According to Kraus et al [1970], "the presence of dental crown abnormalities in four of the 'normal' members of the family suggests that these
individuals were heterozygotes even though the abnormalities were not as
severe or as numerous as those seen in the affected siblings" (page 915).
These authors also suggested that "this may be an instance where the
study of dental crown morphology may reveal the presence of a recessive
mutant in a heterozygous state" (page 915).

An "excess" of affected women (72%) was reported in a review of 46
cases [Taylor, 1957a] but this may be due to chance [Smith, 1970, 1982].
According to Kirkham and Werner [1975], the sex ratio is normal among
affecteds.

A similar condition has been called Werner's syndrome. It consists of
shortness of stature, premature graying of hair, scleropoikiloderma, trophic
ulcers of legs, arteriosclerosis, bilateral juvenile cataract, hoarse and high-
pitched voice, hypogonadism, osteoporosis, tendency to develop diabetes,
and an AR pattern of inheritance. In contrast to Rothmund's syndrome,
patients with Werner's syndrome are essentially normal until the age of 20
or 30, when their hair becomes gray. The skin changes and cataracts
develop after the canities has appeared. Shortness of stature and atrophy
of muscle and subcutaneous fat of the distal extremities are more pronounced in Werner's syndrome.

Other references. Blinstrub et al [1964], Bellafiore et al [1966], Zamith et
al [1974], Kristensen [1975], Marquina et al [1975], Sri-Skanda-Rajah-
Sivayoham and Ratnaike [1975], Gorlin et al [1976], Hall et al [1980].

2. FISCHER-JACOBSEN-CLOUSTON'S SYNDROME (McK 12950)

Synonyms. Hidrotic ectodermal dysplasia; Clouston's syndrome; Wal-
deyer-Fischer's syndrome; Jacobsen's syndrome; Jacobsen-Clouston-

Weech's syndrome; inherited ectodermal dystrophy; ungual type of ecto-dermal dysplasia; hereditary dystrophy of the hair and nails.

Hair. Dry, fine, usually blond; slow-growing; hypotrichosis; absent or scanty eyebrows and lashes.

Teeth. Occasional hypodontia, anodontia, widely spaced teeth, natal teeth, carious teeth. See *Comments.*

Nails. The defects vary considerably and several degrees of dystrophy are frequently found in the same hand; thickened and slightly discolored; striated longitudinally; paronychia; tendency toward horny hypertrophy; convex ends; anonychia.

Sweat. Normal.

Skin. Dry and rough; tendency toward scaliness; hyperpigmentation of some areas; thick dyskeratotic palms and soles.

Hearing. No data.

Eyes. Occasional strabismus, cataract, and myopia.

Face. Normal.

Psychomotor and growth development. Occasional mental deficiency and short stature; speech difficulties.

Limbs. Tufting of terminal phalanges; clubbing of fingers.

Other findings. Thickening of the skull bones.

Etiology. AD.

Comments. Contrary to the opinion of Witkop et al [1975], dental abnor-malities also occur in this condition [Jacobsen, 1928; Clouston, 1929; Aceves-Ortega and Madrigal, 1977]. However, hair and nail defects are the most common and important findings.

Eccrine poromatosis of remarkable severity was reported by Wilkinson et al [1977] in a member of the kindred originally described by Clouston [1929].

Rajagopalan and Tay [1977] described a Chinese family from Malaysia with 15 affected members over five generations. Thus, this condition is not restricted to French-Canadians as supposed by several authors.

Both "hidrotic" and "hydrotic" have been used to name this condition. The first relates to *sweat* and the second to *water* [Taylor, 1957b]. The first is, therefore, the correct one [Freire-Maia and Pinheiro, 1982b].

Case 2 described by Weech [1929] probably presented this condition, and *not* hypodontia and nail dysgenesis [tooth and nail syndrome; Witkop, 1965]. Case 1 probably presented CST syndrome.

Rousset [1952] described six men and one woman over four generations with a dysplasia that is similar to Fischer-Jacobsen-Clouston's syndrome, and may be summarized as follows:

Hair. Trichorrhexis nodosa; hypotrichosis; atrichosis.

Teeth. Widely spaced; pointed canines.

Nails. Thick, brittle, and upturned at the free margins.

This condition seems to be due to an AD gene with incomplete penetrance and variable expressivity.

Other references. Fischer [1910], Clouston [1939], Joachim [1936], Wilkey and Stevenson [1945], Klein [1954], Williams and Fraser [1967], McNaughton et al [1976], Giraud et al [1977], Hazen et al [1980].

3. COFFIN-SIRIS'S SYNDROME (McK 13590)

Synonyms. Mental retardation with absent fifth fingernails and terminal phalanx; syndrome of absent fifth fingernails and toenails, short distal phalanges and lax joints; fifth-digit syndrome.

Hair. Sparse scalp hair; bushy eyebrows and lashes; hirsutism of limbs, forehead, and back.

Teeth. Delayed eruption; microdontia.

Nails. Absent to hypoplastic fifth fingernails and toenails; other nails occasionally hypoplastic or absent.

Sweat. Normal.

Skin. Dermatoglyphic changes; simian crease.

Hearing. No data.

Eyes. Blepharoptosis, hypophoria, hypermetropia, and astigmatism (case 2 of Coffin and Siris [1970]).

Face. Coarse, with thick lips, wide mouth and nose, anteverted nostrils, and low nasal bridge.

Psychomotor and growth development. Retardation of both; hypotonia.

Limbs. Lax joints; clinodactyly of the fifth fingers; general absence of terminal phalanges of fifth fingers and toes; general aplasia or variable hypoplasia of middle and proximal phalanges of other fingers and toes; slender metacarpals and metatarsals; bilateral or unilateral dislocation of the radial heads; small or absent patella.

Other findings. Frequent respiratory infections; umbilical and inguinal hernias; cleft palate; feeding problems in infancy; six lumbar vertebrae; short sternum; microcephaly.

Etiology. AR? AD? See *Comments.*

Comments. Except for the Carey and Hall [1978] report of a brother-and-sister pair from normal, nonconsanguineous parents, all the other cases are sporadic. The fathers of the two girls (cases 1 and 2) described by Coffin and Siris [1970] and of the girl (case 2) described by Weiswasser et

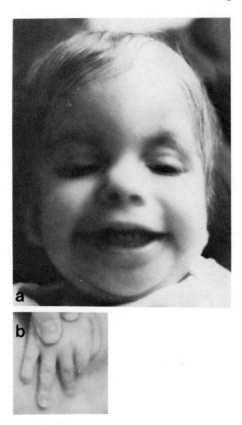

Fig. 4-1. Coffin-Siris's syndrome. a. Hypotrichosis of the scalp, bushy eyebrows and lashes, microcephaly, wide mouth and nose, anteverted nostrils, and microdontia. b. Absence of terminal phalanx in the fifth finger. Courtesy of Dr. John C. Carey, Salt Lake City, UT.

al [1973] were mildly and differently affected (bushy eyebrows, also present in two brothers; ptosis of one eyelid; iris coloboma of one eye; slightly small fingernails and toenails).

Since some authors described sporadic cases and Carey and Hall [1978] reported an affected brother-and-sister pair, this condition—according to the criteria mentioned in Chapter 15—should be assumed to be AR?. However, since mild signs of the condition have been described among close relatives of some propositi, we also prefer to add a possible AD inheritance (AD?).

Weiswasser et al [1973] noted some differences between the cases described by Senior [1971] and those of Coffin and Siris [1970]; they stated that it is "unclear" whether the former patients represent milder variations in the clinical spectrum of the syndrome described in 1970. However, Holmes et al [1972] accepted all of these patients as having the same syndrome.

Mattei et al [1981] identified Coffin-Siris's syndrome with Coffin-Lowry's syndrome (McK 30360); this confusion was entirely due to the fact that Dr. Siris was a coauthor with Dr. Coffin of the description of both syndromes [Coffin et al, 1966; Coffin and Siris, 1970]. Gorlin [1981] noticed this confusion and clarified some basic points regarding nosology in this area: a) They are two different syndromes; b) Coffin-Lowry's syndrome is due to an X-linked gene; c) the cause of Coffin-Siris's syndrome is unknown; it may be heterogenous, with some cases representing examples of the fetal hydantoin syndrome, or due to an AR gene; d) some of the patients described in the literature, such as those of Senior [1971], possibly present a milder form of Coffin-Siris's syndrome.

Other references. Bartsocas and Tsiantos [1970], Feingold [1978].

4. ODONTOTRICHOMELIC SYNDROME (McK 27340)

Synonyms. Tetramelic ectodermal dysplasia; Freire-Maia's syndrome; tetramelic deficiencies, ectodermal dysplasia, deformed ears, and other abnormalities.

Hair. Severe hypotrichosis of scalp and body.

Teeth. Hypodontia; microdontia; coniform teeth; persistence of deciduous teeth.

Nails. Hypoplastic (?). See *Comments.*

Sweat. Normal. See *Comments.*

Skin. Thin, dry, and shiny; an unusual number of wrinkles are formed when the patients smile or grimace; dermatoglyphic disturbances; hypoplastic nipples; hypoplastic or absent areolae.

Hearing. Normal.

Eyes. Normal.

Face. Protruding lips; enlarged nose; large, thin, prominent, and deformed auricles; incomplete right cleft lip in one patient.

Psychomotor and growth development. EEG abnormalities; growth retardation.

Limbs. Extensive tetramelic deficiencies.

Other findings. Metabolic abnormalities (excess of tyrosine and/or triptophane in urine); ECG abnormalities.

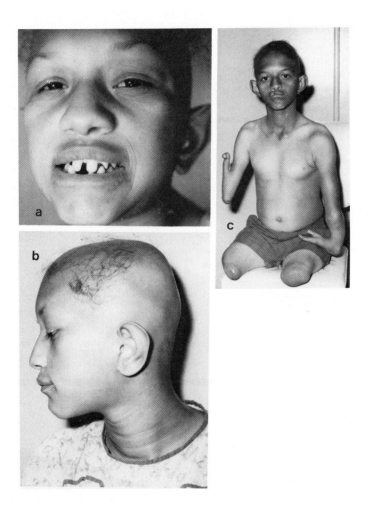

Fig. 4-2.   Odontotrichomelic syndrome. a. Scanty eyelashes, large nose, prominent auricles, hypodontia, peg-shaped and widely spaced teeth, wrinkles beneath the eyes and mouth. b. Hypotrichosis, deformed auricle, and protruding lips. c. Tetramelic limb deficiencies, large nose, protruding lips, incomplete right cleft lip, hypoplastic areolae, and scalp hypotrichosis. Note that a and b are of the same patient. a, c, Freire-Maia [1970a]; b, Chautard and Freire-Maia [1970].

Etiology. AR?

Comments. Freire-Maia [1970a] described one sibship with four affected and four normal persons. Two patients (a boy and a girl) have been examined; the other two died in infancy. Freire-Maia et al [1971] and Cat et al [1972] presented a clinical reappraisal of the patients 2 years after the first analysis. They corrected some information given previously and added some additional data. Detailed sweat analysis of the two patients gave evidence that even though they sweat less than most normal persons, there is no reason to suspect that they have pathologic hypohidrosis [Rapone-Gaidzinski, 1978; see also Alcântara-Silka, 1977]. On the other hand, this condition is a malformation/dysplasia syndrome, not a pure dysplasia. That is why it should be called odontotrichomelic syndrome (OS), and *not* odontotrichomelic hypohidrotic dysplasia.

Nail hypoplasia is doubtful; only one patient had (two) fingers with nails that were abnormally small, but this size reduction could be a secondary effect of the finger malformation.

Brill et al [1972], Pries et al [1974], Bowen and Armstrong [1976], and Gemme et al [1976] identified this condition as the EEC syndrome. However, in spite of some similarities, the differences are sufficiently large to permit considering them different syndromes [Freire-Maia, 1971, 1977a; Preus and Fraser, 1973; Witkop et al, 1975; Pinsky, 1975, 1977; Rapone-Gaidzinski, 1978; Cohen, 1978, 1979]. For a discussion of the problem, see ectrodactyly-ectodermal dysplasia-cleft lip/palate (EEC) syndrome, and Pinheiro and Freire-Maia [1980].

Other references. Freire-Maia et al [1969, 1970], Chautard and Freire-Maia [1970].

5. TRICHODENTOOSSEOUS (TDO) SYNDROME I (McK 13080)

Synonyms. Enamel hypoplasia with curly hair; Robinson-Miller-Worth's syndrome; taurodontism-amelogenesis imperfecta-kinky hair syndrome; AD hypoplastic enamel with hair and nail defect.

Hair. Dry, thick, tough, and with short curls; often straight during childhood; balding may occur with age in men.

Teeth. Small, pitted, and widely spaced; they rapidly become discolored and eroded to the gingival surface (amelogenesis imperfecta); taurodontism; proneness to caries; gingival abscesses and fistulous tracts; patients are edentulous by the second or third decade.

Nails. Flat, thickened, misshapen, and striated; tendency toward peeling and breaking; white bands have been found in some children.

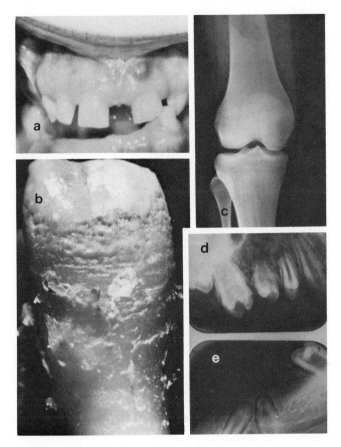

Fig. 4-3. Trichodentoosseous syndrome I. a. Severe hypodontia. b. Pitting of the enamel surfaces. c. Excessive radiodensity of bones. d, e. Taurodontism and impacted permanent teeth. Courtesy of Dr. Ronald J. Jorgenson, Charleston, SC.

Sweat. Normal.
Skin. Occasional lesions.
Hearing. Normal.
Eyes. Normal.
Face. Occasional frontal bossing.
Psychomotor and growth development. Normal.
Limbs. Occasional clinodactyly.
Other findings. Sclerosteosis most evident in the skull, apparently limited to the cortex and unaccompanied by increased cortical thickness (except at

the base of the skull). Some of the calvarial sutures show evidence of premature fusion leading to mild to moderate dolichocephaly.

Etiology. AD. See *Comments*.

Comments. Lichtenstein et al [1972] described a large kindred with 107 affecteds (60 women) over six generations; the name trichodentoosseous syndrome has been proposed by them.

Bone involvement was not detected in the patients of Robinson et al [1966], who seem to be the first authors to describe this syndrome.

The above description refers to the condition Shapiro et al [1983] calls TDO-I. A provisional TDO-II was described by Leisti and Sjöblom [1978] and another possible condition (TDO-III) was proposed by Shapiro et al [1983]. The possibility of variable expressivity of the same gene cannot be ruled out but the fact that the TDO clinical picture seems to fall into three distinct types in different families suggests that they are different genetic entities. The discovery of different types in the same family, the increase of the number of families showing only subtypes II or III, and linkage studies may solve this problem.

Other references. Crawford [1970], Jorgenson and Warson [1973].

6. TRICHODENTOOSSEOUS (TDO) SYNDROME II (McK: not listed)

Synonyms. None.

Hair. Wooly from birth; sparse; easily detachable. Scant or absent axillary and pubic hair. Reduced facial hair growth in men.

Teeth. Enamel hypoplasia; pitting and abscesses; taurodontism; open roots of the incisors; obliteration of the pulp cavities.

Nails. Flat, thin, and brittle.

Sweat. No data.

Skin. No data.

Hearing. No data. See *Other findings*.

Eyes. No data.

Face. Frontal bossing; prognathism.

Psychomotor and growth development. No data.

Limbs. Sclerotic long bones.

Other findings. Thick calvaria; sclerosis of the skull; narrowing of the ear canal in men.

Etiology. AD.

Comments. Leisti and Sjöblom [1978] described four individuals over three generations.

Other reference. Shapiro et al [1983].

7. TRICHODENTOOSSEOUS (TDO) SYNDROME III (McK: not listed)

Synonyms. None.
Hair. Curly; decreased facial hair in the only man examined.
Teeth. Small and widely spaced; pitted dysplastic enamel; abscesses; taurodont molars; short, open roots; impacted teeth; missing left mandibular second bicuspid (one patient); carious teeth (one patient).
Nails. Fingernails are brittle and peeled at their free borders.
Sweat. Normal.
Skin. Normal.
Hearing. Normal.
Eyes. Normal.
Face. Frontal bossing; square jaw.
Psychomotor and growth development. Normal.
Limbs. Undertubulated clavicles.
Other findings. Calvaria: increased density and thickness, obliterated diploë, poorly pneumatized mastoid, obliterated frontal sinuses; chondrocranium: increased density and thickness; macrocephaly.
Etiology. AD.
Comments. Shapiro et al [1983] described 11 persons (five women) over four generations.
Other references. None.

8. INCONTINENTIA PIGMENTI (McK 30830)

Synonyms. Bloch-Sulzberger's syndrome; Bloch-Siemens's incontinentia pigmenti; melanoblastosis cutis linearis sive systematisata; melanosis corii degenerativa; nevus pigmentosus systematicus; Bloch-Sulzberger's melanoblastosis; Siemens-Bloch's pigmented dermatosis.
Hair. Thin; alopecia (pseudopelade) in about one-third of the cases.
Teeth. Hypodontia; anodontia; peg-shaped; delayed eruption. Both deciduous and permanent teeth are affected.
Nails. Dystrophic in all or most of the fingers and toes in about one-tenth of the cases.
Sweat. Normal.
Skin. Vesicular-bullous eruption in the neonatal period followed or accompanied by verrucous lesions and bizarre pigmentation. Pigmented macules may be present at birth.

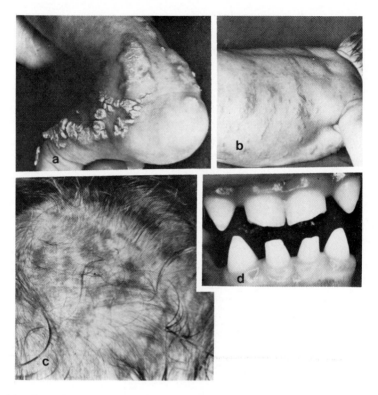

Fig. 4-4. Incontinentia pigmenti. a. Verrucous lesions. b. Erythema and vesicles. c. Hypotrichosis. d. Hypodontia and peg-shaped teeth. a–c, courtesy of Dr. Robert G. Carney, Jr., Decatur, IL; d, courtesy of Dr. Hiroaki Koshiba, Tokyo, Japan.

Hearing. Occasional congenital loss.

Eyes. Ophthalmologic alterations in about one-fifth of the patients include blindness, strabismus, cataract, uveitis, retrolental fibroplasia, optic nerve atrophy, microphthalmia, etc.

Face. Occasional cleft lip.

Psychomotor and growth development. Spastic tetraplegia, hemiplegia, diplegia; epilepsy; mental retardation; occasional short stature. About one-third of the cases present severe CNS anomalies. Hemiatrophy.

Limbs. Occasional clubfoot.

Other findings. Occasional cleft palate; microcephaly.

Etiology. AD (with sex-limited expression); XD or AD (with intrauterine lethality of males). The XD hypothesis seems more probable (see Chapter 16).

Comments. This syndrome is similar to the Naegeli-Franceschetti-Jadassohn's (NFJ) dysplasia, also known as Naegeli's incontinentia pigmenti, which occurs in both sexes, while Bloch-Sulzberger's syndrome (BS) occurs predominantly in women; trichodysplasia and skin inflammation are included in BS but not in NFJ; hypohidrosis and palmoplantar hyperkeratosis are seen in NFJ but not in BS. Naegeli-Franceschetti-Jadassohn's dysplasia will be described later.

Histologically, deposits of melanin are observed in the corium; the designation of the condition was based on the idea that the basal layer of the epidermis is "incontinent" of melanin.

The familial incidence of BS and its almost exclusive occurrence in females (Carney [1976] reported a preponderance of more than 37:1) may be explained by the last two hypotheses above. For discussions of this matter, see especially Curth and Warburton [1965] and Carney [1976].

Other references. Pfeiffer [1960], Fraser and Friedmann [1967], Smith [1969], Carney and Carney [1970], Perlman [1971], Morgan [1971], Piussan et al [1973], Iancu et al [1975], Witkop et al [1975], Berbich et al [1981], Hecht et al [1982].

9. CRANIOECTODERMAL SYNDROME (McK 21833)

Synonyms. Syndrome of skeletal, dental, and hair anomalies; Sensenbrenner-Dorst-Owens's syndrome; cranioectodermal dysplasia; Levin's syndrome.

Hair. Thin, sparse, and slow-growing; abnormal structure (lack of the central pigmented core).

Teeth. Microdontia; hypodontia; widely spaced; enamel hypoplasia; taurodontism; fusion (deciduous teeth).

Nails. Broad and short.

Sweat. Normal.

Skin. Dimples over elbows and knees; bilateral hallucal plantar creases; single flexion crease on each toe; bilateral single palmar creases.

Hearing. No data.

Eyes. Hyperopia; myopia; nystagmus.

Face. Frontal bossing; epicanthal folds and antimongoloid slant; full cheeks; posteriorly angulated pinnae with hypoplastic antihelix; hypotelorism; broad bridge of the nose; anteverted nares; everted lower lip; capillary nevus on the forehead.

Psychomotor and growth development. Hypotonia.

Limbs. Rhizomelic shortness (greatest in the upper limbs); disproportionate shortness of the fibulae; brachychiry; pronounced shortness of middle and distal phalanges of toes and fingers; somewhat flattened epiphyses of the long bones; cutaneous syndactyly; fifth-finger clinodactyly; increased space between the first and second toes; hallux valgus.

Other findings. Dolichocephaly; generalized osteoporosis; highly arched palate; sagittal suture synostosis; short and narrow thorax; pectus excavatum; multiple oral frenula; congenital heart defects (several forms).

Etiology. AR?

Comments. Sensenbrenner et al [1975] described a brother-and-sister pair with a normal sister. Levin et al [1977] reported five children, the first two being those previously described by Sensenbrenner et al (1975), two others being monozygous female twins (with a normal brother) and a sporadic case (a boy, with three normal sibs). All the children had nonrelated parents.

Other reference. Gellis and Feingold [1979].

10. FRIED'S TOOTH AND NAIL SYNDROME (McK: not listed)

Synonyms. None.

Hair. Fine and short; scanty eyebrows.

Teeth. Hypodontia; peg-shaped teeth.

Nails. The fingernails seem to be somewhat thin; the toenails are small, thin, and slightly concave.

Sweat. Normal.

Skin. Normal.

Hearing. Normal.

Eyes. Normal.

Face. Prominent lips and chin; cleft lip (in one patient).

Psychomotor and growth development. Normal.

Limbs. Normal.

Other findings. Branchial cyst on the left side of the neck (in the other patient).

Etiology. AR.

Comments. Fried [1977] described one man and one woman (both from two related sibships with a total of five members and consanguineous parents).

Other references. None.

11. HYPODONTIA AND NAIL DYSGENESIS (McK 18950)

Synonyms. Tooth and nail syndrome; autosomal dominant dysplasia of nails and hypodontia; Witkop's syndrome; Witkop-Weech-Giansanti's syndrome.

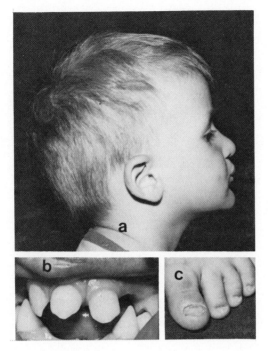

Fig. 4-5.   Fried's tooth and nail syndrome. a. Sparse scalp hair, scanty eyebrows, and prominent lips and chin. b. Hypodontia and peg-shaped teeth. c. Small, thin, and slightly concave toenails. Courtesy of Dr. Kalman Fried, Zerifin, Israel.

Hair. Fine and brittle.

Teeth. Hypodontia; widely spaced; cone-shaped; occasionally supernumerary.

Nails. Small and spoon-shaped; slow growth in children; toenails more severely involved (occasionally at birth); longitudinal ridging.

Sweat. Normal.

Skin. Occasionally dry and somewhat prematurely wrinkled in the face.

Hearing. Normal.

Eyes. Normal.

Face. Everted lips.

Psychomotor and growth development. Normal.

Limbs. Normal.

Other findings. No data.

Etiology. AD.

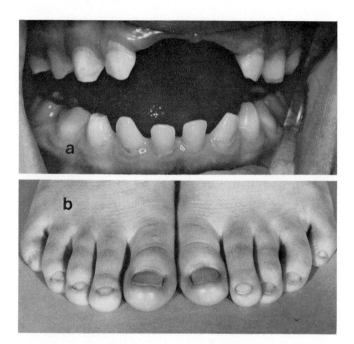

Fig. 4-6. Hypodontia and nail dysgenesis. a. Hypodontia, widely spaced, and cone-shaped teeth. b. Small and spoon-shaped nails. Courtesy of Dr. Carl J. Witkop, Jr., Minneapolis, MN.

Comments. Described by Witkop [1965]. Neither of the two patients described by Weech [1929] seems to have this condition. Patient 1 probably has Christ-Siemens-Touraine's syndrome and patient 2, Fischer-Jacobsen-Clouston's syndrome. Therefore, there seems to be no reason to maintain Weech's name in an eponymic designation of this condition. Hypodontia and onychodysplasia are the only constant signs of this dysplasia [Witkop and Hudson, 1979].

Other references. Giansanti et al [1974], Witkop et al [1975], Hudson and Witkop [1975].

12. DENTOOCULOCUTANEOUS SYNDROME (McK: not listed)

Synonyms. None.

Hair. Scanty body hair with vellus hairs in the moustache and beard areas.

Teeth. Taurodont, pyramidal (single conical root), or fused molar roots (deciduous and permanent teeth).

Nails. Horizontal ridging of the fingernails with distal onychoschizia.

Sweat. Normal.

Skin. Indurated and hyperpigmented over the interphalangeal joints of the fingers (simulation of "knuckle pads").

Hearing. Complete neural loss (unilateral in one patient).

Eyes. Juvenile glaucoma.

Face. Upper lip characterized by absence of "cupid's bow"; thickening and widening of the philtrum; ectropion of both lower eyelids.

Psychomotor and growth development. Normal.

Limbs. Syndactyly (third and fourth fingers); clinodactyly of the fifth finger.

Other findings. No data.

Etiology. AID?

Comments. Ackerman et al [1973a] described one family with nine men and seven women with dental abnormalities over two generations. The other anomalies are mentioned for some of these patients (the propositus and his two affected sibs) without clear reference to the others. The "complete" syndrome is present only in the offspring (two men and one woman) of parents with dental anomalies. This situation looks similar to that observed in XTE syndrome. In the dentooculocutaneous syndrome, it seems that the autosomal gene, in a single dose, produces only dental and lip anomalies, and in the homozygous form it may lead to the "complete" syndrome.

Other reference. Ackerman et al [1973b].

13. TRICHORHINOPHALANGEAL (TRP) SYNDROME I (McK 19035 and 27550)

Synonym. Trichorhinophalangeal dysplasia I.

Hair. Fine, usually blond and sparse (especially in the fronto-temporal areas); slow-growing; medially thick and laterally sparse or absent eyebrows (the Herthoge sign); sparse or normal eyelashes.

Teeth. Occasional supernumerary incisors; microdontia; poorly aligned.

Nails. Occasionally thin, short, and with longitudinal grooves; flattened, koilonychialike and normal in color; racket thumbnails.

Sweat. Normal.

Skin. No data.

Hearing. No data.

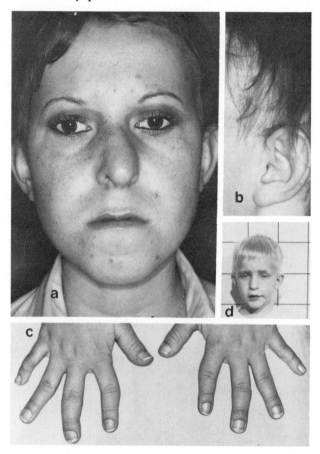

Fig. 4-7.   Trichorhinophalangeal syndrome I. a. Sparse eyebrows, pear-shaped nose, thin upper lip, long philtrum, somewhat large and slightly protruding ears. b. Hypotrichosis at the fronto-temporal area. c. Radial and ulnar clinodactyly; brachydactyly. d. A boy with some manifestations of the syndrome. a–c, courtesy of Dr. Nidia Triay de Juárez, Córdoba, Argentine; d, courtesy of Dr. David L. Rimoin, Torrance, CA.

Eyes. Occasional exotropia and photophobia.

Face. Pear-shaped nose; long and wide philtrum; large and somewhat prominent ears; small protuberance below the lower lip.

Psychomotor and growth development. Short stature.

Limbs. Brachymesophalangy; brachymetacarpy; brachymetatarsy; peripheral dysostosis with type 12 cone-shaped epiphyses [Giedion et al, 1973] at some of the middle phalanges of the hands (the joints are thickened); ulnar and radial deviation of the fingers; occasional clinodactyly; winged scapulae; coxa valga; Perthes-like abnormalities.

Other findings. Narrow palate; scoliosis, lordosis or kyphosis; pectus carinatum; increased susceptibility to upper-respiratory tract infections.

Etiology. AD. Heterogeneity? AR?

Comments. Several examples of AD inheritance are known [Giedion et al, 1973; Felman and Frias, 1977; Peltola and Kuokkanen, 1978; McK 19035], but some cases of affected sibs from normal and apparently non-consanguineous parents were also described [Klingmüller, 1956; Giedion, 1966; McK 27550].

Patients with Langer-Giedion's syndrome (Trichorhinoauriculophalangeal dysplasia or trichorhinophalangeal syndrome II; McK 15023) share many manifestations with TRP syndrome I, but in addition have multiple exostoses, redundant and/or loose skin during infancy, laxity or hypermobility of joints, mild microcephaly, mental deficiency with significantly delayed onset of speech, and skin nevi. Most of the cases with Langer-Giedion's syndrome have been sporadic (Murachi et al [1981] described the first familial case). Several patients present a nonspecific del(8q). It is possible that in patients for whom no deletion could be detected by banding, the condition is due to a minute deletion or to gene mutation [Gorlin et al, 1982].

Other references. Gorlin et al [1969], Weaver et al [1974a], Witkop et al [1975], Pilotto et al [1976], Fukushima et al [1976], Fryns et al [1980, 1981], Zabel and Baumann [1982], Bühler [1982].

14. ELLIS-VAN CREVELD'S SYNDROME (McK 22550)

Synonyms. Chondroectodermal dysplasia; chondrodysplasia ectodermica; chondrodysplasia tridermica; mesoectodermal dysplasia.

Hair. Thin, brittle, sparse, and hypochromic; absent or scanty eyebrows and lashes.

Teeth. Natal teeth; precocious exfoliation; hypodontia (deciduous and permanent teeth); occasionally hypoplastic enamel.

Nails. Dysplastic (brittle, furrowed, and underdeveloped).

Sweat. Normal. See *Comments.*

Skin. Occasional different alterations (hypotrophy, eczema, petechiae, etc) are described in different patients. See *Comments.*

Hearing. Normal.

Eyes. Occasional strabismus, cataract, coloboma of the iris, microphthalmia, exophthalmia, etc. See *Comments.*

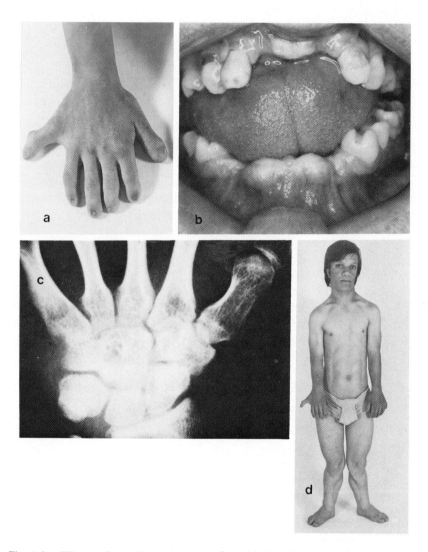

Fig. 4-8.   Ellis-van Creveld's syndrome. a. Complete hexadactyly, rudimentary seventh finger, and severe onychodystrophy. b. Dental abnormalities and gingivolabial fusions. c. Fusion of hamate and capitate. d. Short stature, broad nose, polydactyly, and genua valga. Courtesy of Dr. Rui F. Pilotto, Curitiba, PR, Brazil.

Face. Broad nose; occasional cleft lip, frontal bossing and hypertelorism.

Psychomotor and growth development. Short-limb dwarfism; occasional mild mental retardation. See *Comments.*

Limbs. Bilateral postaxial polydactyly (generally of the hands); brachymetacarpy; thick and short bones of limbs; fusion of the hamate and capitate; clubfoot; genua valga; syndactyly.

Other findings. Congenital heart disease; respiratory difficulties (due to thoracic and tracheobronchial abnormalities); gingivolabial fusion; cleft palate; epispadia; hypospadia; hypoplastic genitalia.

Etiology. AR.

Comments. Four cardinal signs are important for the diagnosis: 1. bilateral postaxial polydactyly; 2. chondrodysplasia of the long bones resulting in acromelic dwarfism; 3. ectodermal defects (generally in the hair, teeth, and nails); 4. congenital heart disease.

According to Pilotto [1978], only 12 of the 204 patients described in the literature (including his own) were analyzed for sweating. Ten of them were euhidrotic. The two patients reported to be hypohidrotic by Lodin and Sjögren [1964] probably did not have the Ellis-van Creveld's syndrome since they did not have polydactyly, fusion of the hamate and capitate, and other typical skeletal abnormalities.

It is doubtful whether skin alterations and mental retardation really represent signs of the syndrome or merely coincidental findings. Skin alterations of different types have been found in only 5 of 204 patients; mental retardation, always mild, was mentioned in only ten patients. There is no eye abnormality (found in 16 patients) that is typical of Ellis-van Creveld's syndrome [Pilotto, 1978].

According to Pilotto [1978], the following abnormalities were found (with their respective frequencies): trichodysplasia (51%), hypodontia of deciduous teeth (88%), hypodontia of permanent teeth (12%), natal teeth (23%), onychodystrophy (93%), congenital heart disease (69%), polydactyly of the hands (96%), polydactyly of the feet (33%), synostosis of hamate-capitate (29%), pseudo cleft lip (13%), and gingivo-labial alterations (55%).

Ellis-van Creveld's syndrome is also classified among the chondrodystrophies or osteochondrodysplasias [Smith, 1969, 1970, 1982; Lamy, 1969; Maroteaux, 1969; Rimoin, 1979]. The largest kindred so far described is that investigated by McKusick et al [1964] in an inbred religious isolate (the Old Order Amish).

Other references. Ellis and van Creveld [1940], Weiss and Crosset [1955], Agostinelli [1970], Pinsky [1977], Waldrigues et al [1977], Silva et al [1980], Rosemberg et al [1983].

15. CYSTIC EYELIDS-PALMOPLANTAR KERATOSIS-HYPODONTIA-HYPOTRICHOSIS (McK: not listed; see *Comments*)

Synonyms. Schöpf-Schulz-Passarge's syndrome; eyelid cysts, hypodontia, and hypotrichosis.

Hair. Generalized hypotrichosis of scalp and body.

Teeth. Extensive hypodontia (rudimentary permanent teeth); persistence of deciduous teeth.

Nails. Brittle with longitudinal and oblique furrows; onycholysis.

Sweat. Normal.

Skin. Palmoplantar keratosis; telangiectatic facial skin; papules; multiple tumors with follicular differentiation.

Hearing. Normal.

Eyes. Bilateral early senile cataract; arteriosclerotic fundi; myopia.

Face. Cysts of eyelids developing late.

Psychomotor and growth development. Normal.

Limbs. Normal.

Other findings. No data.

Etiology. AR.

Comments. Schöpf et al [1971] described two sisters, the offspring of a first-cousin marriage. Burket et al [1983] described one affected man in a noninbred sibship of seven. McKusick [1983] thinks that the condition described by Schöpf et al [1971] is Papillon-Lefèvre's syndrome (McK 24500).

Other references. None.

16. ŠALAMON'S SYNDROME (McK: not listed)

Synonyms. None.

Hair. Dry, inelastic, wirelike, lustreless, and sparse with telogen effluvium in the scalp; pili torti; trichorrhexis nodosa; scanty eyebrows, axillary and pubic hair.

Teeth. Hypodontia; microdontia.

Nails. Highly dystrophic; brittle; ungues plicatae. Toenails are more severely affected.

Sweat. Normal.

Skin. Tendency to develop warts (verruca vulgaris) and papules (verruca plana). Hyperkeratotic scalp.

Hearing. No data.

Eyes. Blepharoconjunctivitis chronica; keratitis punctata; atrophia strati pigmenti retinae; madarosis et trichiasis palpebrae; maculae corneae centrales et periphericae.

Face. Pear-shaped nose.

Psychomotor and growth development. Normal.

Limbs. Slight osteoporosis of long bones.

Other findings. No data.

Etiology. AR.

Comments. Šalamon and Miličevič [1964] described a brother and sister (among seven sibs) from normal consanguineous parents (the degree of consanguinity was unknown). Šalamon et al [1967, 1972] described two sisters (among six sibs) from normal nonconsanguineous parents. The two sibships are apparently unrelated.

Other reference. Šalamon et al [1974].

17. TRICHOOCULODERMOVERTEBRAL SYNDROME (McK: not listed)

Synonyms. None.

Hair. Dry and rough; hypotrichosis of scalp and body; scanty eyebrows (more severely in the distal 1/2 to 2/3) and lashes.

Teeth. Tendency to develop caries; somewhat widely spaced upper incisors.

Nails. Thin and brittle fingernails; wide, short, and dystrophic toenails with paronychia.

Sweat. Normal.

Skin. Dry with tendency to fissures and infection (flexion areas are the most affected); scaling, hyperchromic (ichthyosiformlike) spots on limbs; hyperkeratosis particularly severe on the soles; excess of creases on the digital flexion areas; thick corneal layer; normal sweat glands and absence of sebaceous glands in a small fragment of skin biopsy.

Hearing. Normal.

Eyes. Bilateral nuclear cataract; entropion and trichiasis at the distal 1/3 of the lids.

Face. Characteristic with hypoplasia of nasal alae, prominent lips (a racial trait?), narrow palpebral fissures, slight micrognathia, and prominent bridge of nose.

Psychomotor and growth development. Short stature. See *Comments*.

Limbs. Enlarged interphalangeal joints; apparently hypotrophic phalangeal muscles; partial bilateral cutaneous syndactyly at the base of the fingers (mild webbed fingers); hypoplasia of the palmar eminences; bilateral genu valgum, cubitus valgum, and manus cava.

Other findings. Severe kyphoscoliosis; gingivitis.

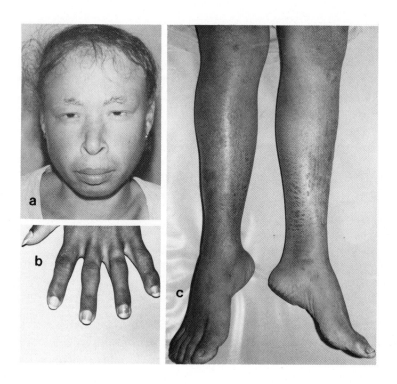

Fig. 4-9.   Trichooculodermovertebral syndrome. a. Scalp hypotrichosis, scanty eyebrows (more severe laterally) and scarce lashes; narrow palpebral fissures; hypoplasia of nasal alae; protruding lips; slight micrognathia; large nose bridge. b. Slight webbing at the bases of the fingers; nails appear normal but are thin and brittle. c. Hyperchromic scaling (ichthyosiformlike) spots; wide, short, and dystrophic toenails. Courtesy of Dr. Auristela F. Paes Alves, Salvador, BA, Brazil.

Etiology. AR.

Comments. Alves et al [1981] described one light mulatto woman, the third member of an inbred (F = 1/64) sibship of four (one man). Short stature may be a familial trait but in this case was also due to kyphoscoliosis.

Other reference. Alves et al [1980].

18. OCULODENTODIGITAL (ODD) SYNDROME II
(McK: not listed)

Synonyms. None
Hair. Mild hypotrichosis.
Teeth. Enamel hypoplasia; microdontia; severe hypodontia; malocclusion; conical mandibular incisors.
Nails. Dysplastic finger- and toenails.
Sweat. No data.
Skin. Normal.
Hearing. No data.
Eyes. Microphthalmia; esotropia; nystagmus; partial atrophy of the iris.
Face. Slight anteversion of the nostrils; small alae nasi; microstomia with furrowing and overlap of the upper lip; micrognathia.
Psychomotor and growth development. Normal.
Limbs. Polydactyly of the left hand; absent terminal phalanx of the right second digit; camptodactyly of the left fifth finger; long thumbs; absent middle phalanx and ulnar deviation of the right fifth finger; syndactyly of the first and second, and of the third and fourth right toes; rudimentary extra toe on each foot.
Other findings. No data.
Etiology. Unknown.
Comments. O'Rourk and Bravos [1969] described a boy belonging to a sibship of three (two women) from normal nonconsanguineous parents.
Other references. None.

19. ARTHROGRYPOSIS AND ECTODERMAL DYSPLASIA
(McK: not listed)

Synonyms. None.
Hair. Hypotrichosis of scalp (atrichia at birth) and body; scanty eyebrows and lashes. Hair, with a nonkinky thin shaft, is twisted about its own axis (stretched pili torti) and the cross-section is oval.
Teeth. Enamel hypoplasia.
Nails. Absent at birth; later they are of normal length; tendency toward longitudinal breaks.
Sweat. Normal.
Skin. Dry; tendency to excessive bruising and scarring after injuries and scratching.

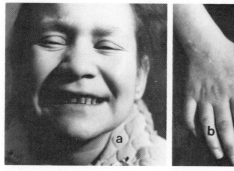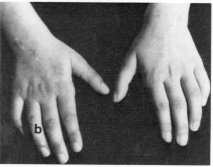

Fig. 4-10.   Arthrogryposis and ectodermal dysplasia. a. Scalp hypotrichosis (not clearly seen), scanty eyebrows and lashes; enamel hypoplasia. b. Maximum extension of fingers; clinodactyly. Courtesy of Dr. Gilbert B. Côté, Athens, Greece.

Hearing. Normal. (The patient produces large amounts of very dry and dark ear wax).

Eyes. Normal.

Face. Bilateral epicanthus; slight mongoloid slant.

Psychomotor and growth development. Short stature (120 cm.; well below the third centile); probable low–normal intelligence level.

Limbs. Bilateral clinodactyly; slight bilateral syndactyly of the second and third toes; slight right talipes equinovarus.

Other findings. Arthrogryposis of all joints (more severe in hands; the patient walks like a very old lady and cannot squat without support); diabetes mellitus; no menarche at the age of 16.

Etiology. Unknown.

Comments. Côté et al [1982] described a 16-year-old girl. She had two younger sibs (a brother and a sister). The normal parents deny consanguinity although they were born in the same small Greek village.

Other references. None.

20. TRICHOODONTOONYCHODERMAL SYNDROME (McK: not listed)

Synonyms. None.

Hair. Hypotrichosis in the parieto-occipital region; area of alopecia in the centro-parietal region due to aplasia cutis congenita; scanty eyelashes without tendency to curl upwards; sparse and irregular eyebrows, especially at their outer 2/3.

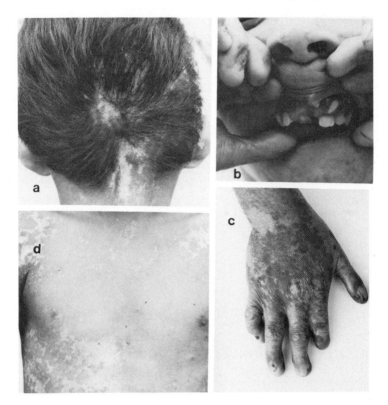

Fig. 4-11.   Trichoodontoonychodermal syndrome. a. Hypotrichosis. b. Enamel hypoplasia and supernumerary teeth. c. Syndactyly, clinodactyly, hypoplastic and flexionless right thumb, short second finger (due to hypoplasia of middle and distal phalanges) and dystrophic nails. d. Absent right nipple; irregular areolae; hypochromic, atrophic and poikilodermalike spots. a–c, unpublished photos; d, Pinheiro et al [1981b].

Teeth. Hypodontia; persistence of deciduous teeth; enamel hypoplasia; delayed eruption; abnormal shape and size; supernumerary teeth.

Nails. Severely dystrophic or absent finger- and toenails; two normal fingernails.

Sweat. Normal.

Skin. Dry with hypochromic, atrophic, and poikilodermalike spots (some with telangiectasis); absent right nipple; irregular areolae; wrinkled back of hands; palmar keratosis; dermatoglyphic alterations (mosaic patchy distribution of dermal ridges both in the palms and in the fingertips); excess of "white lines"; bilateral transverse palmar crease; skin biopsies from the thorax and nose revealed focal areas of hyperkeratosis and large amounts of pigment in the epidermis (vicarious hyperchromia); aplasia cutis congenita of the scalp.

Hearing. Normal.

Eyes. Leukoma in the right eye.

Face. Long philtrum; microstomia with thin lips; hypoplastic right alae nasi; hyperpigmented eyelids and periorbital regions; right palpebral ptosis; mild micrognathia; "cup" ears.

Psychomotor and growth development. Normal.

Limbs. Bilateral clinodactyly of the fifth fingers; syndactyly of the second and third fingers of both hands, of the second and third toes of both feet, and of the fourth and fifth left toes; hypoplastic and flexionless right thumb; pronounced manus cava; hypoplastic distal and middle phalanges of both second fingers; absent middle phalanges of all toes.

Other findings. Mild asymmetry of the skull; congenital hypertrophy of the frenum linguae.

Etiology. Unknown.

Comments. Pinheiro et al [1981b] described a 12-year-old boy from normal, apparently nonconsanguineous parents.

Other reference. Pinheiro et al [1980].

21. TRICHOODONTOONYCHIAL DYSPLASIA (McK: not listed)

Synonyms. None.

Hair. Extensive area of alopecia totalis on the top of the head with only a peripheral fringe of hair on the temporal and occipital regions, where there is dry, brittle, and scarce hair; sparse eyebrows and scanty eyelashes.

Teeth. Enamel hypoplasia in both dentitions; secondary anodontia.

Nails. Variable degree of dystrophy of finger- and toenails.

Sweat. Normal.

Skin. Increased number of pigmented nevi (some at the epidermis level and others forming papules); extranumerary nipples; keratotic actinic lesions, crusts, and ephelides in the scalp; mild palmoplantar hyperkeratosis; dermatoglyphic alterations.

Hearing. One patient had a mixed mild hearing deficit on the left (her external auditory meatus ended blindly). See *Comments.*

Eyes. Normal.

Face. Normal.

Psychomotor and growth development. Short stature.

Limbs. Normal.

Other findings. None.

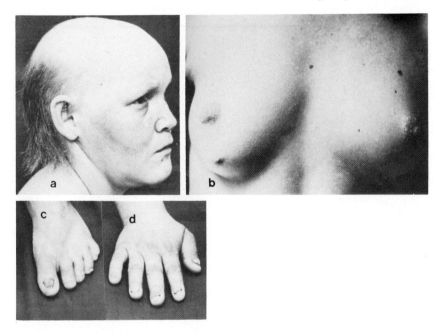

Fig. 4-12.   Trichoodontoonychial dysplasia. a. Alopecia with hypotrichosis of the peripheral fringe, eyebrows, and lashes. b. This patient has two supernumerary nipples, but only
the right one, which is larger and has an areola, is seen in the photo. In a and b, note the
presence of nevi. c, d. Dystrophic nails. a, c, and d show one patient; b, another one.
Unpublished photos.

Etiology. AR?

Comments. Pinheiro et al [1983c] described four sisters belonging to a
noninbred sibship of 13 individuals (eight women). The corrected segregation ratio (with exclusion of the proposita) (9:3) corresponds exactly to a
3:1 ratio. The fact that only women are affected in a sibship of eight
women and five men may be accepted as coincidental.

Unilateral hearing deficit verified in only one patient may be coincidental.

Other reference. Pinheiro and Freire-Maia [1981a].

22. ODONTOONYCHODYSPLASIA WITH ALOPECIA
(McK: not listed)

Synonyms. None.

Hair. Almost total alopecia; absence of eyebrows and lashes, and of
axillary and pubic hair.

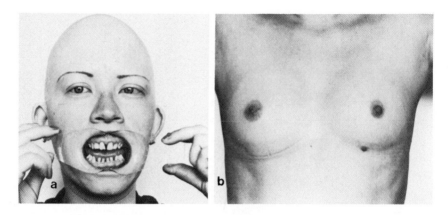

Fig. 4-13.   Odontoonychodysplasia with alopecia. a. Generalized hypotrichosis (eyebrows are painted), enamel hypoplasia, microdontia (both upper central incisors are repaired), and abnormal auricles. b. Left areola with irregular margins; two Montgomery glands at the right areola and one supernumerary nipple below the left breast; axillary atrichosis. Unpublished photos.

Teeth. Enamel hypoplasia; abnormal shape; microdontia; widely spaced.

Nails. Mildly dystrophic. Brittle fingernails, with tendency toward secondary infection. Thick and spoon-shaped toenails. Bilateral subungual corneal layer in the first and second fingers of one patient.

Sweat. Normal.

Skin. Mild palmoplantar keratosis (with scaliness in one patient); supernumerary nipple; irregular areolae; hypertrophied Montgomery glands; dermatoglyphics: absence of digital triradius b, and normal finger patterns; simian creases.

Hearing. Normal.

Eyes. Unilateral deviation of the lacrimal duct (in one patient; at left). Myopia and left esotropia in one patient. See *Comments.*

Face. Mild "cup" ears in one patient.

Psychomotor and growth development. Normal.

Limbs. Bilateral webbed toes (one-third of second and third toes; see *Comments*). Mild webbing at the basis of the fingers.

Other findings. The eldest patient had an ovarian cyst that was surgically removed.

Etiology. AR?

Comments. Pinheiro and Freire-Maia [1981b] described two sisters, the only children of a possibly consanguineous normal couple. Esotropia is not a sign of the condition, since the patient's mother and maternal grandfather had bilateral esotropia.

It is possible that webbing of toes also is not a sign of the condition, since its prevalence is relatively high.

Gorlin and Červenka [1975] described a 26-year-old girl who combined some traits of Turner's syndrome with signs of ectodermal dysplasia. She proved to be a XO/XX mosaic. The signs are the following: alopecia totalis, nail dysgenesis (finger- and toenails), amelogenesis imperfecta (both dentitions), bilateral ptosis of the upper eyelids, small and infantile nipples, absence of breast tissue, absence of the labia majora and minora, no subcutaneous fat in the pubic area, hypertrophic clitoris, no vaginal opening, absence of the vestibule of the vagina, agenesis of the distal one-third of the vagina with a rigid imperforate hymen, hypoplastic cervix, schizoid personality, and low intelligence level (IQ equal to 86). (Sweat and height are normal). Such a combination of signs has never been described before. The relationship of the clinical findings with the chromosome status is not known. No family data are provided. This combination of signs (even excluding those due to Turner mosaicism) is different from that we report here. At the subject index, this "condition" will appear as alopecia totalis, nail dysplasia, and amelogenesis imperfecta.

Other reference. Pinheiro et al [1984]

23. SCHINZEL-GIEDION'S SYNDROME (McK 26915)

Synonym. Schinzel-Giedion's midface-retraction syndrome.

Hair. Generalized hypertrichosis (4/4).

Teeth. Delayed eruption is reported in two patients at 10 and 16.5 months.

Nails. Narrow, deeply set, triangular (2/4) and hyperconvex (3/4).

Sweat. No data.

Skin. Abundant on the neck (4/4); hypoplastic nipples (2/4); hypoplastic dermal ridges (3/4); simian creases (3/4).

Hearing. Normal.

Eyes. Normal.

Face. Saddle nose with depressed root and short bridge (4/4); high and protruding forehead (4/4); orbital hypertelorism (4/4); small and malformed auricles (4/4); anteverted nostrils (4/4); midface hypoplasia (4/4); facial hemangiomata (2/4).

Psychomotor and growth development. Severe mental retardation (3/4); growth retardation (3/4); abnormal EEG and seizures (3/4); spasticity (2/4); recurrent apneic spells (4/4).

Limbs. Mesomelic brachymelia (3/4); postaxial hexadactyly (right foot and left hand) (1/4); hypoplasia of distal phalanges in hands and feet (4/4); short metacarpals of thumbs (2/4); talipes (3/4). See *Other findings.*

Other findings. Multiple bone anomalies were detected radiologically: short and sclerotic base of skull (3/4), multiple Wormian bones (4/4), wide cranial sutures and fontanelles (4/4), broad ribs (3/4), broad cortex and increased density of long tubular bones and vertebrae (2/4), hypoplastic or aplastic pubic bones (2/4); etc. Choanal stenosis (2/4; a third patient is "snuffly"); congenital hydronephrosis (3/4); short and broad neck (3/4); atrial septal defect (2/4); short penis with hypospadia (2/2); deep sulcus between major and minor labia (2/2); hymenal atresia and short perineum (1/2); highly arched palate (2/4); macroglossia (1/4).

Etiology. AR?

Comments. Schinzel and Giedion [1978] described a brother-and-sister pair, Donnai and Harris [1979] described a boy, and Kelly et al [1982] described a girl. All parents were nonconsanguineous and normal.

Other references. None.

24. GROWTH RETARDATION-ALOPECIA-PSEUDOANO-DONTIA-OPTIC ATROPHY (GAPO) (McK: not listed)

Synonyms. Pseudoanodontia, growth retardation and alopecia; pseudoanodontia, cranial deformity, blindness, alopecia, and dwarfism, alopecia, anodontia and cutis laxa.

Hair. Generalized atrichia (some hair, present at birth, is lost during infancy).

Teeth. Pseudoanodontia (all buds are present) of both primary and permanent dentition with absence of alveolar ridges.

Nails. Hyperconvexity on fingers and toes is referred in two patients.

Sweat. Normal.

Skin. Dry; redundant; fragile with inadequate wound-healing (small, depressed scars); depigmented areas; unusual wrinkles; leatherlike and thick on nape and upper back; abnormal dermatoglyphics are referred in two patients.

Hearing. Sensorineural hypoacusia.

Eyes. Optic atrophy; glaucoma; keratoconus; nystagmus; photophobia.

Face. "Small" and "characteristic"; asymmetric. Craniofacial dysostosis; micrognathia; protruding and thickened lips; protruding auricles; prominent supraorbital ridges; depressed nasal bridge; minor malformations in the auricles.

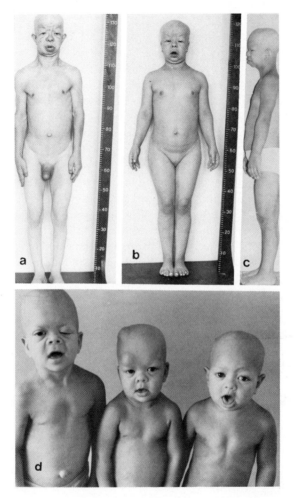

Fig. 4-14.   Growth retardation-alopecia-pseudoanodontia-optic atrophy. a, b, c. Short stature, generalized atrichia, hypoplasia of mammary glands (the third patient is also a female), general appearance of premature aging, and characteristic face. d. Three brothers; note pectus excavatum in the three, umbilical hernia in the first (present but not seen in the other two), and sequels of a neurosurgery performed to alleviate a suspected (but doubtful) intracranial hypertension in the second. a, b, courtesy of Dr. Anita Wajntal, São Paulo, SP, Brazil; c, courtesy of Dr. Elias O. da Silva, Recife, PE, Brazil; d, courtesy of Dr. Antonio Ricardo T. Gagliardi, Brasilia, DF, Brazil.

Psychomotor and growth development. Dwarfism; occasional mental retardation.

Limbs. Symmetrical proximal shortening of humeri; hyperextensible fingers; second and third toes smaller than the fourth; mild brachychiry; wide gap between hallux and second toes.

Other findings. Wide anterior fontanel; prominent scalp veins; increased aminoaciduria; hepatosplenomegaly; bilateral choanal atresia; hyperplasia of sublingual connective tissue; hypoplasia of mammary glands; pectus excavatum; umbilical hernia; delayed bone maturation through childhood and adolescence.

Etiology. AR.

Comments. Andersen and Pindborg [1947] described a noninbred isolated girl; Epps et al [1977] a boy and a girl in an inbred sibship of nine; Fuks et al [1978] a girl in an inbred sibship of four. Wajntal et al [1982] redescribed the patients of Epps et al [1977]. Shapira et al [1982] redescribed the patients of Fuks et al [1978], but the coefficient of inbreeding of the sibship given by Fuks et al [1978] seems to be the correct one (3/16 instead of 1/8). Gorlin [1982a] described a girl in a noninbred sibship of three, and Silva [1982, pers. comm.] described a girl in an apparently noninbred sibship of five (another affected sib died before ascertainment of the family). Gagliardi et al [1983] described three boys, the only children of a normal consanguineous couple. Tipton and Gorlin [1983] redescribed the patient of Gorlin [1982a] and reviewed the literature.

This progerialike condition has been diagnosed as Rothmund-Thomson in the 1977 paper but duly corrected by Wajntal et al [1982].

Other references. Wajntal [1981, pers. comm.], Gorlin [1982, pers. comm.], Gagliardi [1983, pers. comm.].

25. ECTODERMAL DYSPLASIA WITH SYNDACTYLY
(McK: not listed)

Synonym. Ectodermodysplasia syndrome with pillous anomaly and syndactyly.

Hair. Hypotrichosis. Scalp hair is brittle, and either dark and thick or blond and fine; pili torti; sparse eyebrows and lashes.

Teeth. Severe crown hypoplasia; delayed and atypical eruption of permanent teeth.

Nails. Yellowish and partially thickened.

Sweat. Normal.

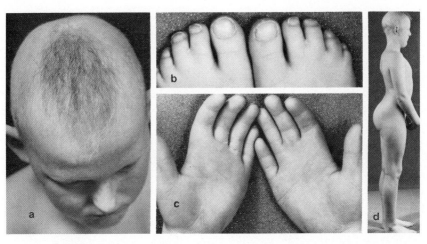

Fig. 4-15.   Ectodermal dysplasia with syndactyly. a. Hypotrichosis of scalp and eyebrows. b. Syndactyly and dystrophic toenails. c. Syndactyly, dystrophic fingernails, hyperkeratotic palms, and simian crease. d. Hypotrichosis and lordosis. Courtesy of Dr. H.-R. Wiedemann, Kiel, West Germany.

Skin. Dry, with hyperkeratosis, especially at the distal third of the trunk, lower limbs, and palmoplantar regions (axillae and elbows are normal). Transverse crease on both palms.

Hearing. Normal.

Eyes. Mild crowding of the lenses; discrete hypermetropia.

Face. Normal.

Psychomotor and growth development. Normal.

Limbs. Syndactyly on both fingers and toes to variable degrees.

Other findings. Lordosis; highly arched palate.

Etiology. AR.

Comments. Wiedemann et al [1978] described three affected children (two boys) belonging to an inbred (F = 1/16) sibship of five.

Other reference. Wiedemann [1982, pers. comm.].

26. OSTEOSCLEROSIS AND ECTODERMAL DYSPLASIA (McK: not listed)

Synonyms. None.

Hair. Short; breaks easily; ribbonlike and straight.

Teeth. Caried and misaligned.

Nails. Split on all fingers and toes.

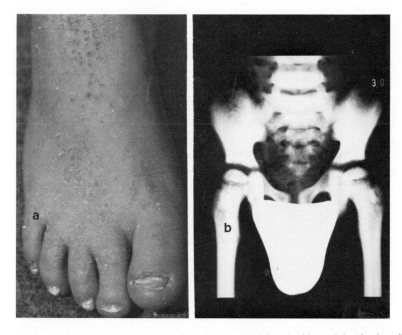

Fig. 4-16.   Osteosclerosis and ectodermal dysplasia. a. Ichthyotic skin and dysplastic nails.
b. Osteosclerosis. Courtesy of Dr. Gilbert B. Côté, Athens, Greece.

Sweat. Normal.
Skin. Ichthyotic and highly sun-sensitive.
Hearing. Normal.
Eyes. Normal.
Face. Normal.
Psychomotor and growth development. Low–normal intelligence; short stature.
Limbs. Clinodactyly on the left fifth finger. See *Comments.*
Other findings. Osteosclerosis (first noticed at age 4); large and prominent papillae on tongue; chronic leukopenia; immunoglobulins with inconsistent patterns of abnormalities; lymphocyte metaphases show an increased frequency of sister chromatid exchanges; left cryptorchidism.
Etiology. AR.
Comments. Côté and Katsantoni [1982] described a boy aged 8 years and 117 cm tall (below the third centile). His birthweight was 2,100 gm. The parents are third cousins and also have a normal daughter. Both parents have cylindrical hair but the mother's is sparse and falls off easily; the father's baldness started at the age of 20. The maternal grandmother (not belonging to the consanguinity line) lost most of her hair permanently at

the age of 50. Clinodactyly is present in the maternal grandmother and in the patient's father.

Other reference. Côté [1982].

27. DERMOODONTODYSPLASIA (McK: not listed)

Synonyms. None.

Hair. Dry; slow-growing (scalp, moustache, and beard); thin moustache; a circumscribed area of alopecia at the top of the head; normal eyebrows and lashes; sparse axillary and pubic hair; normally distributed in the rest of the body.

Teeth. Hypodontia; microdontia; persistence of deciduous teeth.

Nails. Dysplastic; brittle (on fingers and toes).

Sweat. Normal.

Skin. Dry and thin to variable degree (especially on palmoplantar regions); simian crease.

Hearing. Normal.

Eyes. Normal. See *Comments*.

Face. Left palpebral ptosis; prognathic mandible. See *Comments*.

Psychomotor and growth development. Normal.

Limbs. Normal.

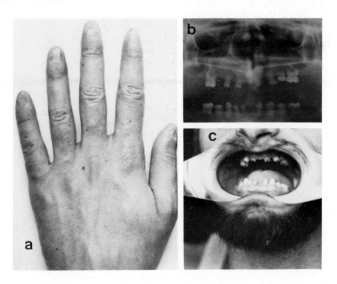

Fig. 4-17.   Dermoodontodysplasia. a. Dystrophic fingernails. b, c. Dental defects. a, c, unpublished photos; b, Pinheiro and Freire-Maia [1983].

Other findings. None.

Etiology. AD.

Comments. A kindred with 11 affected persons (seven women) in four generations was described by Pinheiro and Freire-Maia [1983]. The propositus is the most affected of all and the only with trichodysplasia. The other members of the family present variable but always mild combinations of the clinical signs listed above. Skin alterations are the most frequent sign, dental anomalies being second.

Bilateral myopia and astigmatism have been detected in the propositus, his mother, maternal grandfather, and paternal greatgrandfather; since the last two persons do not present the dysplasia picture, this eye involvement is not a part of the condition.

Prognathic mandible was also seen in the propositus's maternal grandmother, also showing the dysplasia.

Other reference. Pinheiro and Freire-Maia [1982].

28. TRICHOODONTOONYCHODYSPLASIA WITH PILI TORTI (McK: not listed)

Synonyms. None.

Hair. Pili torti; sparse blond scalp hair.

Teeth. Hypodontia; widely spaced teeth; peg-shaped incisors; abnormal eruption.

Nails. Dysplastic, flat fingernails with linear grooves.

Sweat. Normal. See *Comments*.

Skin. Pale and mildly dry.

Hearing. Normal.

Eyes. Normal.

Face. Mild maxillary hypoplasia; thinness of lip.

Psychomotor and growth development. Normal.

Limbs. Normal.

Other findings. No data.

Etiology. AD? XD?

Comments. Carey [1982, pers. comm.] described a 4-year-old girl, the only child of nonconsanguineous parents. Her mother presents peg-shaped permanent maxillary right and left incisors and hypodontia. This may represent a mild manifestation of an autosomal or X-linked gene.

The patient may be mildly hypohidrotic, but this does not seem to be a pathologic trait; she does not have fever episodes.

This condition shares some traits with pili torti and enamel hypoplasia (McK 26190).

Other references. None.

29. MESOMELIC DWARFISM-SKELETAL ABNORMALITIES-ECTODERMAL DYSPLASIA (McK: not listed)

Synonyms. None.
Hair. Hypotrichosis.
Teeth. Dysmorphic; irregular eruption; malpositioned. See *Comments.*
Nails. Hypoplastic toenails.
Sweat. Normal.
Skin. Extremely hypoplastic papillary dermal ridges; bilateral transitional palmar flexion creases.

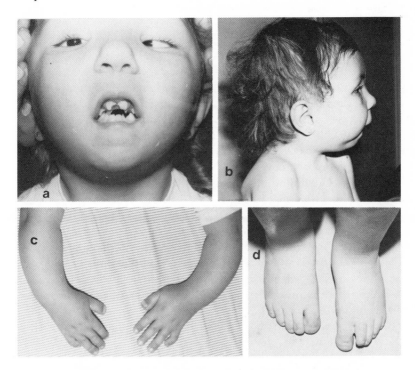

Fig. 4-18.   Mesomelic dwarfism-skeletal abnormalities-ectodermal dysplasia. a. Esotropia, antimongoloid palpebral slanting, epicanthal folds, depressed nasal root, long philtrum, malpositioned and malformed teeth. b. Hypotrichosis and micrognathia. c. Short and curved forearms, short hands, duplicated thumbs (with nail duplication), and camptodactyly. d. Hypoplastic toenails and broad halluces. Courtesy of Dr. Décio Brunoni, São Paulo, SP, Brazil.

Hearing. Normal.

Eyes. Esotropia.

Face. Depressed nasal root; micrognathia; hypertelorism; antimongoloid palpebral slanting; epicanthal folds; long philtrum; thin lips.

Psychomotor and growth development. Short stature; mild psychomotor retardation. See *Comments.*

Limbs. Short forearms and hands; agenesis of the proximal third of the right radius and of the proximal two-thirds of the left radius; broad thumbs; proximal placement of thumbs with duplication of the distal phalanx and nail duplication; brachymesophalangy and camptodactyly of both fifth fingers; short legs; flattened acetabular margins; broad halluces.

Other findings. Brachycephaly; narrow and highly arched palate; retarded ossification of the anterior fontanel with the presence of Wormian bones; not-aired facial sinuses; large mandibular angle.

Etiology. Unknown.

Comments. Brunoni et al [1982] described an affected girl, aged 4 years, the only daughter of normal nonconsanguineous parents. The family history was unremarkable. She was born weighing 2,800 gm after an 8-month first pregnancy by cesarean section for fetal distress. Cyanotic, she needed special care for 3 days. At infancy she had many episodes of catarrhal bronchitis. Until 24 months she had a slow psychomotor development but a later (October 1982) psychometric evaluation (Gesell test) was normal. However, her growth remained below 2 SD.

Cytogenetic G banding, cardiological, hematological, and immunoglobulins evaluations were normal.

The syndromes that more closely resemble the present condition are Aarskog's syndrome (probably due to an X-linked recessive gene; McK 30540) and Robinow's syndrome (perhaps a case of heterogeneity; AD; AR? McK 18070). However, hypotrichosis, nail dystrophy, digital anomalies, highly arched palate, brachycephaly, antimongoloid palpebral slanting, etc, seen in the present patient, do not seem to have been reported in Robinow's syndrome.

Other references. Brunoni [1982, pers. comm.], Brunoni et al [1984].

30. ECTODERMAL DYSPLASIA SYNDROME WITH TETRAMELIC DEFICIENCIES (McK: not listed)

Synonyms. None.

Hair. Scalp hypotrichosis; sparse eyebrows and lashes.

Teeth. Hypodontia; peg-shaped teeth; widely spaced.

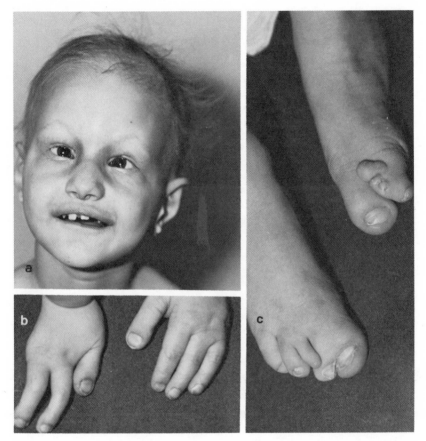

Fig. 4-19.   Ectodermal dysplasia syndrome with tetramelic deficiencies. a. Scalp hypotrichosis, sparse eyebrows and lashes, wide nasal root, esotropia, hypodontia, cone-shaped teeth, and large ears. b. Extensive bone deficiencies and platonychia. c. Extensive bone deficiencies; dysplastic toenails. Courtesy of Drs. A. Schinzel and A. Klingenberg, Zürich/ St. Gallen, Switzerland.

Nails. Platonychia; mildly dysplastic toenails.
Sweat. Normal.
Skin. Hypoplastic areolae and nipples.
Hearing. Normal.
Eyes. Esotropia.
Face. Malformed auricles; protruding lips; wide nasal root.
Psychomotor and growth development. Probably mild mental retardation; short stature.

Limbs. Bipartite right clavicle; short and distally curved left clavicle; coxa valga; steep femoral necks; short and bent right radius; extremely hypoplastic right ulna; absence of fourth and fifth right fingers; absence of the fifth left ray with only one carpal bone; short metacarpal and distal phalanx of the left thumb (this phalanx is bifid). Left foot is highly malformed; absence of fourth and fifth rays; duplication of terminal phalanx of the third ray; right foot: synostosis of cuboid and lateral cuneiform bones; cutaneous syndactyly between first and second, and (partially) between third and fourth toes; absence of the third metatarsal bone; thin proximal phalanx of the third toe.

Other findings. Microcephaly; constant tearing and repeated infections of the conjunctivae (atresia of the naso-lacrimal ducts?).

Etiology. Unknown.

Comments. Schinzel and Klingenberg [1981, pers. comm.] described a 2-year-old boy from normal nonconsanguineous parents. The propositus has a younger normal sib.

Other references. None.

5. Subgroup 1-2-4

1. REGIONAL ECTODERMAL DYSPLASIA WITH TOTAL BILATERAL CLEFT (McK: not listed)

Synonyms. None.

Hair. Scarce and only present on the skull circumference ("the rest of the head is quite hair-less"; page 100); eyebrows and lashes consist of short, thin and very light colored single hairs.

Teeth. Anodontia.

Nails. Normal.

Sweat. Anhidrosis on the head; defective sweat glands.

Skin. Thin, shiny, desquamating on the entire head, with dermoid cysts.

Hearing. No data.

Eyes. Ectropion; epidermalization on the conjunctiva of everted right upper lid; hypoplastic tarsal plates; bilateral lagophthalmos during sleep.

Face. Bilateral cleft lip; aplasia of the premaxilla.

Psychomotor and growth development. Normal.

Limbs. Normal.

Other findings. Bilateral cleft palate.

Etiology. Unknown.

Comments. Fára [1971] described a girl, the only child from nonconsanguineous parents.

Other references. None.

2. MELANOLEUCODERMA (McK 24650)

Synonyms. Berlin's syndrome; Berlin's ectodermal dysplasia; leucomelanoderma, infantilism, mental retardation, hypodontia, hypotrichosis; congenital generalized melanoleucoderma associated with hypodontia, hypotrichosis, stunted growth, and mental retardation.

Hair. Dry and abundant scalp hair showing a slight tendency toward premature grayness; sparse eyebrows with absence of the lateral parts;

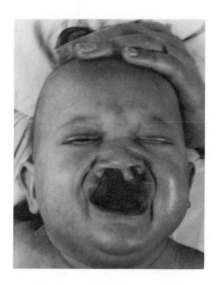

Fig. 5-1. Regional ectodermal dysplasia with total bilateral cleft.   Severe hypotrichosis of scalp, eyebrows, and lashes. Severe bilateral cleft lip and palate; anodontia. Courtesy of Dr. M. Fára, Prague, Czechoslovakia.

normal eyelashes; axillary and pubic hair almost normal in the women and absent in the men; absence of lanugo hair; small moustache (similar in both sexes).

Teeth. Delayed eruption of deciduous and permanent teeth; hypodontia.

Nails. Normal.

Sweat. Mild hyperhidrosis of the palms and soles was mentioned in two of four patients. See *Comments*.

Skin. Pale, dry, thin, smooth, pliable, and feminine; generalized mottled dyschromia consisting of various shades of hyper- and hypopigmentation; anetopoikilodermalike lesions over the elbows, knees, and proximal phalangeal articulations; pyoderma over the lower regions of the legs leading to atrophic scars; palmoplantar hyperkeratosis.

Hearing. Normal.

Eyes. Myopia and slight strabismus (one patient).

Face. Typical "family face" with flat saddle nose, thick lips with slight telangiectasia, and deep furrows around eyes and mouth.

Psychomotor and growth development. Mental retardation; short stature.

Limbs. Moderate hyperextensibility of the fingers; slender legs.

Other findings. Sexual underdevelopment in men (hypospadias, small penis and scrotum, atrophy of the testes, absence of secondary sexual characteristics).

Etiology. AR.

Comments. Berlin [1961] described four (two women) of 12 sibs from normal first-cousin parents. Sweating was described as normal although mild palmoplantar hyperhidrosis was mentioned in association with hyperkeratosis in two of the four cases.

Other references. None.

3. BÖÖK'S DYSPLASIA (McK 11230)

Synonyms. Böök's syndrome; premolar aplasia-hyperhidrosis-canities prematura (PHC) syndrome; premolar aplasia-hyperhidrosis-premature greying.

Hair. Canities prematura.

Teeth. Hypodontia of the premolar region (absence of one to eight teeth); secondary backward displacement of the canines and persistence of the deciduous premolars.

Nails. Normal.

Sweat. Palmoplantar hyperhidrosis.

Skin. Normal.

Hearing. No data.

Eyes. Blue irides. See *Comments.*

Face. No data.

Psychomotor and growth development. Normal.

Limbs. No data.

Other findings. No data.

Etiology. AD.

Comments. Böök [1950] described one kindred with about 300 members (172 included in the pedigree). Sixty-three persons (including 25 affected) were submitted to clinical and odontological examinations; however, the clinical analysis was based on 18 completely examined individuals. The corrected segregation ratio among the children of matings involving an affected spouse was 24A:19N, which corresponds to 1:1. Penetrance is complete but expressivity is variable.

Blue irides were observed in all 18, but they are also a very common trait in Sweden, hence they may be a coincidence. However, considering that these patients have deficient production of hair pigments, a connection with iris pigmentation may be possible.

Other references. None.

4. CONGENITAL INSENSITIVITY TO PAIN WITH ANHIDROSIS (McK 25680)

Synonyms. Congenital sensory neuropathy with anhidrosis; congenital familial sensory neuropathy with anhidrosis.

Hair. Hypotrichosis in areas of the scalp.

Teeth. Enamel aplasia.

Nails. See *Comments*.

Sweat. Hypohidrosis with hyperthermia. See *Comments.*

Skin. Dry; scars from self-inflicted bites may be present on the fingers and arms; chronic sores are common on the hands, feet, and pressure points, such as the buttocks.

Hearing. No data.

Eyes. Irregular lacrimation.

Face. See *Other findings.*

Psychomotor and growth development. Mental retardation.

Limbs. Multiple fractures from trauma resulting in deformities; the fractures are not noticed and, therefore, the injured areas are not adequately treated; joint degeneration (Charcot joints).

Other findings. Universal sensory loss; absent pain perception and physiologic responses to painful stimuli; impaired temperature and touch perception; diminished tendon reflexes; occasional encopresis and enuresis; ulceration of the mouth and scars from biting tongue and lips.

Etiology. AR.

Comments. The presence of histologically normal sweat glands suggests a defect in the neuro-effector mechanism. Local sweating was not induced either by electrical stimulation or by intradermal injection of pilocarpine and other substances. However, Vassella et al [1968] obtained local sweating in their patient by simultaneous application of acetylcholine and adrenaline.

Only one patient (among a total of eight) was described as having fractured and dystrophic nails by Pinsky and DiGeorge [1966], but it is doubtful that this is a sign of the syndrome. However, if onychodysplasia proves to be a part of it, the condition should be moved to the 1-2-3-4 subgroup.

The number of familial cases is (2+2)/7 and of consanguinity (uncle-niece marriage) is 1/7 [Vassella et al, 1968].

There is a condition, first described by Dearborn [1932]—congenital analgesia (also AR)—without ectodermal signs [reviews and new data in Beçak et al, 1964; Saldanha et al, 1964].

Other references. Swanson [1963].

5. LENZ-PASSARGE'S DYSPLASIA (McK: not listed)

Synonym. Lenz's dysplasia.

Hair. Hypotrichosis (precocious baldness among men, and less severe, although variable, manifestation among women); trichorrhexis fissurata; abnormal distribution of body hair.

Teeth. Variable degrees of hypodontia (more severe among men).

Nails. Normal.

Sweat. Hypohidrosis; lower number of sweat pores.

Skin. Increased incidence of nonspecific lesions.

Hearing. Normal.

Eyes. Normal.

Face. Normal.

Psychomotor and growth development. Normal.

Limbs. Normal.

Other findings. No data.

Etiology. XD.

Comments. Lenz [1963] described mild hypotrichosis, hypodontia, normal nails, and euhidrosis in one family with seven affected persons (two men) over three generations. This family was reinvestigated by Passarge [1972, 1980, pers. communications] who described two other patients and found hypohidrosis as a new sign of the condition. Settineri et al [1976] described another family with 74 affected persons (41 men) but gave them the diagnosis of Christ-Siemens-Touraine's syndrome "with some unusual features" (page 212) [see discussions in Freire-Maia, 1977c; Pinheiro, 1977]. A detailed pedigree of this large Brazilian kindred was presented by Settineri [1964]. The penetrance of the gene was estimated to be about 0.85.

There is another condition called Lenz's syndrome (Lenz's microphthalmia; microphthalmia, and digital anomalies; microphthalmia or anophthalmos with associated anomalies) due to an X-linked recessive gene (McK 30980).

Other references. None.

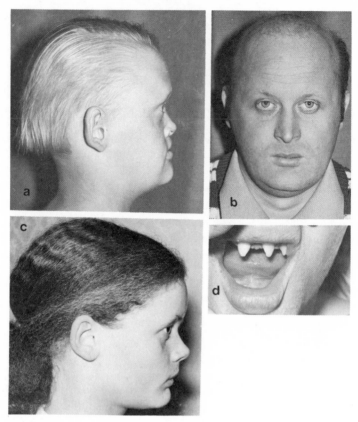

Fig. 5-2. Lenz-Passarge's dysplasia.   a. Hypotrichosis. b. Alopecia. c. Curly hair reflecting structural change. d. Severe hypodontia in a man. Courtesy of Dr. Eberhard Passarge, Essen, West Germany.

6. ECTODERMAL DYSPLASIA WITH PALATAL PARALYSIS (McK: not listed)

Synonym. Congenital ectodermal dysplasia, anhidrotic, with palatal paralysis and associated chromosome abnormality.

Hair. Absence of frontal hair, eyebrows, and lashes.

Teeth. Stunted and peg-shaped; enamel hypoplasia.

Nails. No data.

Sweat. Anhidrosis on face (absent sweat glands).

Skin. Absence of sebaceous glands on the face.

Hearing. Conductive loss; otitis.

Eyes. Normal.

Face. Frontal bossing; depressed nasal bridge.

Psychomotor and growth development. Normal.

Limbs. Normal.

Other findings. Highly arched palate; palatal paralysis; diminished sensation in the palate, the posterior pharyngeal wall and the tonsillar pillar area; abnormal and distorted speech with a marked nasal component.

Etiology: Unknown.

Comments. Wesser and Vistnes [1969] described one girl (in a sibship of five) from normal nonconsanguineous parents.

Fifteen cells studied showed normal complement of 46 chromosomes and normal female sex chromatin, but eight of them presented a consistent abnormality: "an incomplete break at the junction of the middle and distal third of one of the limbs" of a chromosome of Group A (page 397). No further details or photos were provided.

Other references. None.

6. Subgroup 1-3-4

1. FISCHER'S SYNDROME (McK: not listed)

Synonym. Fischer-Volavsek's syndrome.
Hair. Sparse scalp hair, eyebrows, and lashes.
Teeth. Normal.
Nails. Onychogryposis; onycholysis.
Sweat. Palmoplantar hyperhidrosis.
Skin. Occasional xeroderma; palmoplantar keratosis.
Hearing. No data.
Eyes. No data.
Face. Edema of the eyelids.
Psychomotor and growth development. Occasional mental deficiency.
Limbs. Clubbing of the distal phalanges of the fingers and toes.
Other findings. Syringomielia; apathy.
Etiology. AD.
Comments. Fischer [1921] described eight affected individuals (seven men) over five generations.
Other references. Volavsek [1941], Gorlin et al [1964].

2. TRICHODYSPLASIA-ONYCHOGRYPOSIS-HYPOHIDROSIS-CATARACT (McK: not listed)

Synonyms. Freire-Maia's syndrome; hypohidrotic ectodermal dysplasia (Freire-Maia type).
Hair. Straight, brittle, and dry, with marked frontal upsweep; partial alopecia with follicular hyperkeratosis of scalp; scanty eyebrows and lashes.
Teeth. Normal.
Nails. Severe onychogryposis (hands and feet).
Sweat. Hypohidrosis (suggestive of a reduced number of sweat glands).
Skin. Dry and warm, with slight follicular hyperkeratosis; hyper- and hypochromic spots on limbs and face; dermatoglyphic alterations.

117

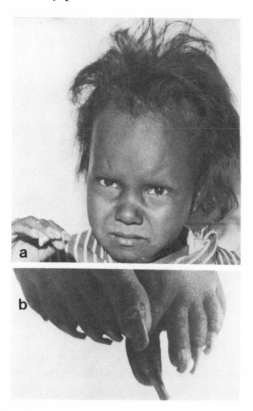

Fig. 6-1. Trichodysplasia-onychogryposis-hypohidrosis-cataract. a. Straight, dry, and brittle hair with frontal upsweep; scanty eyebrows and lashes; saddle nose; note dystrophic fingernails. b. Severe onychogryposis [Freire-Maia et al, 1975].

Hearing. Normal.

Eyes. Bilateral nuclear cataract.

Face. Frontal bossing; depressed nasal bridge.

Psychomotor and growth development. Both retarded.

Limbs. Normal.

Other findings. None.

Etiology. Unknown.

Comments. Freire-Maia et al [1975] described one girl whose family could not be located. The above eponymic designation should not be used since there is another condition with the same eponym (the odontotrichomelic syndrome).

Other reference. Smith and Knudson [1977].

3. ALOPECIA-ONYCHODYSPLASIA-HYPOHIDROSIS-DEAFNESS (McK: not listed)

Synonyms. None.

Hair. Extensive hypotrichosis with few, sparse, thin, yellowish scalp hair; absence of eyebrows; sparse lashes.

Teeth. Normal.

Nails. Normal fingernails; thick, slightly deformed and darkish toenails, with subungual hyperkeratosis and irregularities in the free margins; anonychia at birth.

Sweat. Hypohidrosis as determined by the pilocarpine iontophoresis test and by application of 0-phthaldialdehyde in xylene.

Skin. Hyperpigmented (with a tan color); dry and slightly rough, with hyperkeratosis of palms, soles, knees, and elbows; dermatoglyphics with extensive ridge dissociation.

Hearing. Sensorineural deafness.

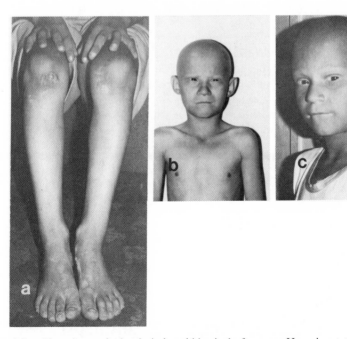

Fig. 6-2. Alopecia-onychodysplasia-hypohidrosis-deafness. a. Hyperkeratotic lesions on knees and feet; dysplastic toenails. b, c. Alopecia, absent eyebrows, esotropia, mongoloid slant of palpebral fissures, narrow palpebral fissures, anteverted auricles with broad upper antihelical region, and pectus excavatum [Freire-Maia et al, 1977].

Eyes. Bilateral esotropia; photophobia; severe hyperopia.

Face. Unusual, with prominent nose and nose bridge; slightly anteverted auricles with broad upper antihelical region; mongoloid palpebral slanting and narrow palpebral fissures.

Psychomotor and growth development. EEG abnormalities without seizures; short stature.

Limbs. Normal.

Other findings. Pectus excavatum; retarded bone age.

Etiology. Unknown.

Comments. Freire-Maia et al [1977] described one girl (with a normal brother) from normal nonconsanguineous parents. Her short stature may be a familial trait.

Other reference. Freire-Maia et al [1981].

4. HAYDEN'S SYNDROME (McK: not listed)

Synonyms. None.

Hair. No scalp hair, eyebrows, or lashes; virtually no body hair.

Teeth. Normal.

Nails. Severe pachyonychia (hands and feet).

Sweat. Severe hypohidrosis.

Skin. Follicular and placquelike hyperkeratosis; almost ichthyosislike hyperkeratosis on the shins; extremely severe palmoplantar hyperkeratosis to the point of almost complete stiffness of the fingers and toes; severe chronic scalp infection with many pustules.

Hearing. Chronic external otitis of such severity that no one has ever seen the tympanic membrane; it thus leads to virtual deafness.

Eyes. Severe chronic conjunctivitis leading to virtual blindness.

Face. Saddle nose; narrow palpebral fissures.

Psychomotor and growth development. Normal.

Limbs. No data.

Other findings. No data.

Etiology. Unknown.

Comments. Dr. Patricia W. Hayden at the University of Washington, Seattle [Opitz, 1975; Cohen, 1980] studied one patient from normal non-consanguineous parents.

Other references. None.

5. ALOPECIA-ONYCHODYSPLASIA-HYPOHIDROSIS (McK: not listed)

Synonyms. None.

Hair. Absent scalp and body hair; no eyebrows or lashes.

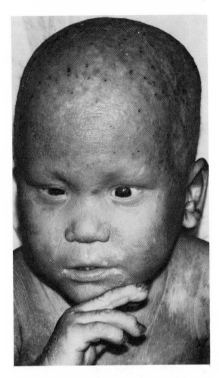

Fig. 6-3. Hayden's syndrome. Generalized hyperkeratosis and scalp infection with pustules; saddle nose; narrow palpebral fissures; pachyonychia; generalized hypotrichosis. Courtesy of Dr. M. Michael Cohen, Jr., Seattle, WA.

Teeth. Normal.

Nails. Severely dystrophic (thick and yellow).

Sweat. Hypohidrosis with hyperthermia.

Skin. Thick, scaly skin in patches over most of the body (the scalp, soles, and legs are more severely affected); eczema; scaly lesions with crusting and some open sores most pronounced around orifices.

Hearing. Normal.

Eyes. Photophobia; horizontal nystagmus; legal blindness.

Face. Large and prominent ears; protruding lips; saddle nose.

Psychomotor and growth development. Short stature; IQ around 38; seizures (as many as four per day when not on medication).

Limbs. Normal.

Other findings. Hypospadia; nonpalpable testes (normal size penis, capable of erection); generalized marginal gingivitis.

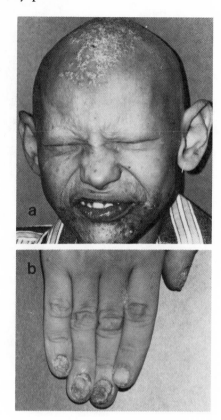

Fig. 6-4.   Alopecia-onychodysplasia-hypohidrosis. a. Large and prominent auricles, protruding lips, photophobia, scaling lesions with crusts, and open sores. b. Severely dystrophic (thick and yellow) nails. Courtesy of Dr. Lawrence H. Shire, Columbia, SC.

Etiology. Unknown.

Comments. Shire [1979, pers. comm.] described a boy among three sibs from normal nonconsanguineous parents. The patient had corneal transplants at 9 years of age.

Other references. None.

6. HYPOHIDROTIC ECTODERMAL DYSPLASIA WITH HYPOTHYROIDISM (McK 22505)

Synonyms. Ectodermal dysplasia, hypohidrotic, with hypothyroidism and ciliary dyskinesia; HEDH syndrome.

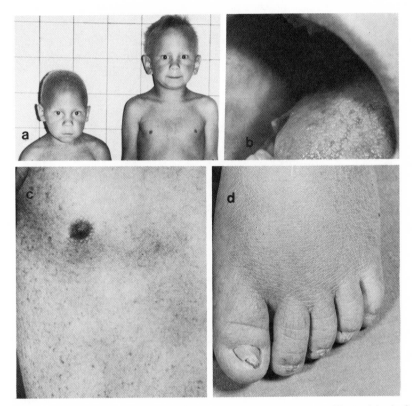

Fig. 6-5. Hypohidrotic ectodermal dysplasia with hypothyroidism. a. Hypotrichosis. b. Buccal pigmentation. c. Pigmented spots. d. Severe nail dystrophy. Courtesy of Dr. H.F. Pabst, Edmonton, Alberta, Canada, and Dr. O. Groth, Linköping, Sweden.

Hair. Scanty and wispy scalp hair with a hard, haylike consistency; scanty eyebrows and normal lashes.

Teeth. Normal.

Nails. Dystrophic, ridged finger- and toenails with a shriveled appearance.

Sweat. Hypohidrosis with hyperthermia; low number of sweat gland pores in the palms.

Skin. Poorly developed palmar dermal ridges; mottled brownish skin pigmentation of the trunk during the first months of life; in one patient at the age of 5, the speckled appearance of the skin began to fade gradually, leaving face and distal limbs looking lightly tanned, while the dark brown freckles of the trunk faded more slowly into a light café-au-lait tone; no change of pigmentation occurred in the other patient; marked dermatographia.

Hearing. Normal. Ear canals usually filled with dry crumbly cerumen.

Eyes. Tear ducts frequently blocked with resultant bilateral epiphora; frequent conjunctivitis.

Face. Normal.

Psychomotor and growth development. Short stature.

Limbs. Normal.

Other findings. Retarded bone age; lacy pigmentation of the buccal mucosae; structural ciliary abnormalities of the respiratory tract; recurrent and severe upper and lower respiratory tract infections (until 5 years of age); severe cow's milk intolerance in infancy; elevated thyrotrophin; decreased thyroid hormone production from early childhood; no evidence of thyroid tissue shown by radiolabeled iodine studies. See *Comments.*

Etiology. AR?

Comments. Pabst et al [1981] described two brothers among three sibs from normal nonconsanguineous parents. The recurrent respiratory infections may have the ciliary abnormalities as a contributing factor.

Other references. Pabst and Groth [1975], Pabst [1979, pers. comm.].

7. ECTODERMAL DYSPLASIA WITH SEVERE MENTAL RETARDATION (McK: not listed)

Synonyms. None.

Hair. Absent scalp (except for a small wisp in the center of the head) and body hair.

Teeth. Normal.

Nails. Almost absent from fingers and toes, being represented by a thin and soft membrane on the first and second digits.

Sweat. Hypohidrosis without hyperthermia.

Skin. Fine, thin, and shiny with some desquamation over the hands, feet, and top of the head; absence of both nipples and of one areola.

Hearing. Normal.

Eyes. Blindness with bilateral cataract.

Face. Abnormal ears.

Psychomotor and growth development. Severely retarded. The patient never walked, talked (occasionally emits some sounds), or was toilet trained.

Limbs. Normal.

Other findings. Absence of menstruation; prepubertal vulva.

Etiology. Unknown.

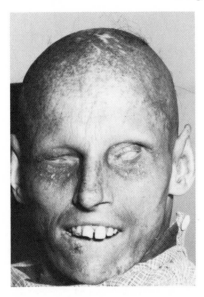

Fig. 6-6. Ectodermal dysplasia with severe mental retardation. Severe hypotrichosis (scalp, eyebrows, and lashes), blindness (bilateral cataract), abnormal auricles, and skin desquamation. Courtesy of Dr. S. Shivanathan, Carshalton, Great Britain.

Comments. Kirman [1955] described one woman among six sibs from normal nonconsanguineous parents. She was born in 1936; the photograph was taken in 1971.

Other reference. Kirman [1980, pers. comm.].

8. ALOPECIA UNIVERSALIS-ONYCHODYSTROPHY-TOTAL VITILIGO (McK: not listed)

Synonyms. Total vitiligo, total alopecia and nail changes.

Hair. Progressive loss of body and scalp hair, eyebrows, and lashes within a brief period of time. Loss of the protective nasal hair and eyelashes leads to chronic nasal congestion and conjunctivitis.

Teeth. Normal.

Nails. Dystrophic fingernails and toenails with transverse ridging, proximal thickening, distal desquamation, friability, and opacity. The dystrophic changes vary from mild to severe in the same individual.

Sweat. The patients complain of an increased tendency to sweat. No test was applied to evaluate this hyperhidrosis.

Skin. Total vitiligo (a complete, rapid, and uniform loss of pigment without going through a patchy state). The cutaneous surface becomes light and translucent, and prone to sunburns.

Hearing. Normal.

Eyes. Normal.

Face. Normal.

Psychomotor and growth development. Normal.

Limbs. Normal.

Other findings. No data.

Etiology. AR?

Comments. Lerner [1961] described three sporadic patients. Hair, skin, sweat, and nail changes occurred gradually within a short period of time after the onset of hypotrichosis at the respective ages of 13, 37, and 39 years. Demis and Weiner [1963] described two other sporadic cases with onset of manifestations at 23 and 46 years, respectively. This seems to be the only ectodermal dysplasia whose signs are all noncongenital.

Other references. None.

9. DERMOTRICHIC SYNDROME (McK: not listed)

Synonyms. None.

Hair. Generalized atrichia from birth.

Teeth. Normal.

Nails. Dystrophic and hyperconvex fingernails.

Sweat. Hypohidrosis without hyperthermia.

Skin. Generalized ichthyosiform lesions all over the body, including palmoplantar areas and scalp.

Hearing. No data.

Eyes. No data.

Face. Prominent forehead; large ears; small nose with mildly low nasal bridge; blepharophimosis.

Psychomotor and growth development. Severe psychomotor retardation; abnormal EEG; frequent apyretic seizures; short stature.

Limbs. Normal.

Other findings. Hemivertebrae at the dorsolumbar region; congenital aganglionic megacolon (surgery at 12 months); narrow arched palate; positive Benedict and glucoseoxydase tests; discrete increase of tyrosinemia; discrete anemia.

Etiology. XR.

Comments. Silva [1982, pers. comm.] reported six affected men over two generations. The above data were obtained for the propositus, at the age of 3 years, the only one alive at the time.

Other references. None.

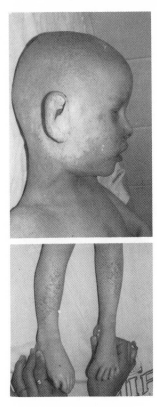

Fig. 6-7. Dermotrichic syndrome. Generalized atrichia, large ears, small nose with low nasal bridge, blepharophimosis, and generalized ichthyosiform lesions. Courtesy of Dr. Elias O. da Silva, Recife, PE, Brazil.

7. Subgroup 2-3-4

1. AMELOONYCHOHYPOHIDROTIC DYSPLASIA (McK 10457)

Synonyms. Hypoplastic enamel-onycholysis-hypohidrosis; ameloonycho-hypohidrotic syndrome.

Hair. Normal.

Teeth. Hypocalcified-hypoplastic enamel that is brown, pitted, thin, and soft; multiple unerupted permanent teeth that undergo resorption.

Nails. Onycholysis involving 1/4 to 1/2 of the distal portion of the finger- and toenails, with subungual hyperkeratosis.

Sweat. Hypofunction of sweat glands.

Skin. Generally xerotic with keratosis pilaris over the buttocks and extensor surfaces of the limbs; seborrheic dermatitis of scalp.

Hearing. No data.

Eyes. No data.

Face. No data.

Psychomotor and growth development. No data.

Limbs. No data.

Other findings. No data.

Etiology. AD.

Comments. Witkop et al [1975] described one family with 11 affected (eight men) over three generations. Witkop and Sauk [1976] redescribed this kindred and reported that, after the publication of the 1975 paper, another family was observed with the same condition.

Other references. None.

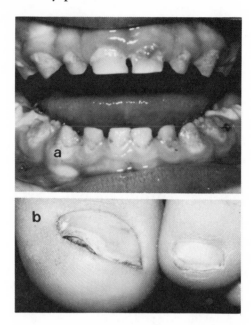

Fig. 7-1. Ameloonychohypohidrotic dysplasia. a. Primary dentition showing hypoplastic and pitted enamel (it was soft and brittle on many teeth). b. Subungual hyperkeratosis. Courtesy of Dr. Carl J. Witkop, Jr., Minneapolis, Minnesota.

8. Subgroup 1-2

1. OROFACIODIGITAL (OFD) SYNDROME I (McK 31120)

Synonyms. Papillon-Léage's and Psaume's syndrome; Gorlin-Psaume's syndrome; orodigitofacial dysostosis; oralfacialdigital syndrome I; dysplasia linguofacialis.

Hair. Dryness and/or variable degree of alopecia (65%).

Teeth. Absence of the lower lateral incisors (50%); malposition; occasional supernumerary canines and enamel hypoplasia.

Nails. Normal.

Sweat. Normal.

Skin. Evanescent facial milia (100%).

Hearing. Normal.

Eyes. Dystopia canthorum.

Face. Broad nasal root (95%); hypoplasia of alae nasi (70%); median cleft of the upper lip (40%). Occasional short philtrum, frontal bossing, and ear abnormalities.

Psychomotor and growth development. Occasional mental retardation (usually mild), trembling, and short stature.

Limbs. Several types of malformations such as brachydactyly, clinodactyly, syndactyly, polydactyly, etc (90%); occasional genu valgum.

Other findings. Multiple hypertrophied lingual and labial frenula (ca 100%); lateral grooving of maxillary alveolar process (90%); grooved anterior alveolar process of mandible (60%); lobate tongue (ca 100%); ankyloglossia (30%); cleft palate (80%); hypoplasia of malar bone (75%). Hypoplasia of the base of the skull, renal abnormalities, hydrocephaly, etc, have also been described in some patients.

Etiology. XD (lethal in men).

Comments. Gorlin and Psaume [1962] studied 22 cases, reviewed the literature, and stated that OFD I occurs in approximately 1% of the patients with cleft palate.

131

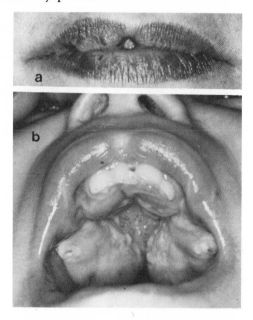

Fig. 8-1. Orofaciodigital syndrome I. a. Median cleft of the upper lip. b. Hypodontia, malposition, lateral grooving of maxillary alveolus; note also central cleft of upper lip. Courtesy of Dr. Robert J. Gorlin, Minneapolis, MN.

Rimoin and Edgerton [1967] suggested that Mohr's syndrome should be called OFD syndrome II (McK 25210). In addition to the different mode of inheritance (Mohr's syndrome is AR), this condition has none of the skin and hair changes of the X-linked OFD syndrome I, but shows conductive hearing defect and bilateral polysyndactyly of halluces, not present in OFD syndrome I. Note that there is also an OFD syndrome III [Sugarman et al, 1971; McK 25885]. Neither OFD syndrome II or OFD syndrome III is an ectodermal dysplasia of Group A [Freire-Maia 1971, 1977a].

Hooft and Jongbloet [1964] referred to two brothers in an inbred sibship of five, whose parents were normal. Besides the classical facial changes, one of the patients also had bilateral microphthalmia and unilateral coloboma of the optic papilla. This may be another condition.

Other references. Wahrman et al [1966], Aita [1969], Whelan et al [1975], Harrod et al [1976], Gorlin [1979].

2. OCULODENTODIGITAL (ODD) SYNDROME I (McK 16420)

Synonyms. Oculodentoosseous dysplasia; Meyer-Schwickerath's and Weyer's syndrome; Gillespie's syndrome; dysplasia oculodentodigitalis;

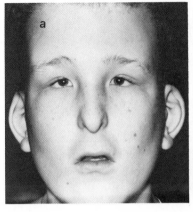 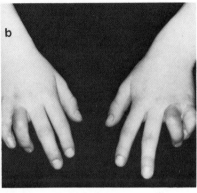

Fig. 8-2. Oculodentodigital syndrome I. a. Abnormal auricles, hypotelorism, sparse eyebrows and lashes, microphthalmia, reduced lid apertures, epicanthal folds, hypoplastic nasal alae. b. Syndactyly (surgically repaired), clinodactyly, and camptodactyly. Courtesy of Dr. Robert J. Gorlin, Minneapolis, MN.

microphthalmos syndrome; microphthalmos, enamel hypoplasia, and camptodactyly of fifth fingers syndrome; oculodentodigital dysplasia.

Hair. Brittle, sparse, and dry; slow-growing.

Teeth. Generalized hypoplasia of the enamel; occasional hypodontia, microdontia, and premature loss.

Nails. Normal.

Sweat. Normal.

Skin. Normal.

Hearing. Occasional conductive impairment.

Eyes. Microcornea; microphthalmia with small orbits; reduced lid apertures. Occasional findings: optic atrophy, synechiae, disc coloboma, persistence of pupillary membrane, nystagmus, congenital cataract, glaucoma, strabismus, and epicanthal folds. See *Comments.*

Face. Characterized by a thin nose, hypoplastic alae, and narrow nostrils; cleft lip; orbital hypotelorism; occasional micrognathia and mild pinna defects. See *Comments.*

Psychomotor and growth development. Occasional mild mental retardation.

Limbs. Syndactyly and camptodactyly of the fourth and fifth fingers; aplasia or hypoplasia of the middle phalanges of the fifth fingers and of one or more toes; occasional ulnar clinodactyly of the fifth fingers and syndactyly of the third and fourth toes; hip dislocation; cubitus valgus.

Other findings. Microcephaly; cranial hyperostosis; cleft palate; osteopetrosis.

Etiology. Heterogeneity. AD and AR. See *Comments.*

Comments. The original description was presented by Lohmann in 1920, but Meyer-Schwickerath, Grüterich, and Weyers, in 1957, delineated the condition and gave it the name oculodentodigital dysplasia [Vittori and Carbonnel, 1976].

Since hypotelorism, hypertelorism, and normal distance between the orbits have been referred in patients with ODD syndrome I, Fára and Gorlin [1981] investigated the problem in 12 patients and found orbital (bony) hypotelorism in five and normal interorbital distance among the other seven. These authors also verified a medial shift of the orbits, leading to normal inner intercanthal distance and to decreased outer intercanthal distance. The length of the palpebral slids is always decreased.

An AD gene is generally accepted as the cause of this syndrome. However, Taysi et al [1971] stated that among the 21 families found in the literature, the inheritance pattern was clearly that of an AD gene in only six. All the other cases were sporadic, some with associated parental consanguinity. The Turkish authors pointed out that we may be dealing with at least one of the following situations: 1. Different, but very similar, syndromes (with different causes); 2. All cases are due to an AD gene with incomplete penetrance and variable expressivity. The father of one of their patients had a thin nose with hypoplastic alae, suggesting weak expressivity of the gene. Other sporadic cases may be due to fresh mutation; 3. The sporadic cases may be phenocopies.

Due to the fact that there are situations showing a clear AD pattern of inheritance and others with isolated cases associated with parental consanguinity, we prefer to assume that ODD syndrome I represents a group of at least two different conditions (see below), one of which would be due to an AR mechanism of inheritance.

Dr. J.H. Renwick, in 1967, found that a constant feature of this condition is the absence of the middle phalanx of the toes (2nd through 5th) [personal communication to McKusick, 1979, 1983].

O'Rourk and Bravos [1969] described a sporadic case with some manifestations of ODD syndrome but with differences that led them to assume that they were dealing with a new syndrome (ODD syndrome II). This patient had dysplastic nails, unilateral preaxial polydactyly, and absence of the terminal phalanx of the right second digit and of the middle phalanx of the fifth right finger. ODD syndrome II was mentioned in the 1-2-3 subgroup.

Beighton et al [1979] described one isolated case from normal nonconsanguineous parents and two patients from two related and possibly inbred sibships that presented ODD syndrome with some distinctive changes:

marked cranial hyperostosis, massive mandibular overgrowth, gross clavicular widening and serious neurological complications. These patients may either represent a new syndrome (in which case perhaps at least four conditions may have been described as ODD) or an extreme expression (characterized by the severity of skeletal involvement and neurological complications) of the same recessive gene as in one of the ODD syndromes.

According to the above analysis, the four possible ODD syndromes are ODD syndrome I (AD), ODD syndrome I (AR), ODD syndrome II (unknown cause), and Beighton's ODD syndrome (AR?).

Other references. Gorlin et al [1963], Reisner et al [1969], Thodén et al [1977].

3. HALLERMANN-STREIFF'S SYNDROME (McK 23410)

Synonyms. Hallermann's syndrome; Hallermann-Streiff-François' syndrome; François' dyscephaly; dyscephalic syndrome of François; François syndrome; Ullrich and Fremery-Dohna's syndrome; dyscephalia mandi-

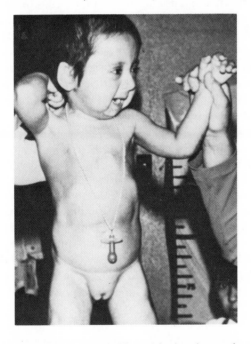

Fig. 8-3. Hallermann-Streiff's syndrome. Hypotrichosis, abnormal auricle, thin and pointed nose, micrognathia. Courtesy of Dr. Oswaldo Frota-Pessoa, São Paulo, SP, Brazil.

bulooculofacialis; dyscephaly with congenital cataract and hypotrichosis; mandibulofacial dysmorphia; mandibulooculofacial dysmorphism; oculomandibulofacial syndrome; oculomandibulodyscephaly with hypotrichosis.

Hair. Thin and light; generalized or sutural alopecia of scalp.

Teeth. Natal; supernumerary; hypodontia; malocclusion with open bite; persistence of deciduous; premature caries; coniform teeth; hypoplastic enamel.

Nails. Normal. See *Comments.*

Sweat. Normal.

Skin. Cutaneous atrophy largely limited to the face and/or scalp.

Hearing. Normal.

Eyes. Bilateral microphthalmia; congenital cataract; congenital corectopia; occasional nystagmus, strabismus, blue sclerae, optic disc coloboma, various chorioretinal pigment alterations, etc.

Face. Characteristically birdlike (thin, small, and pointed nose, with hypoplasia of the cartilage, becoming parrotlike with age); the head has an abnormal shape, usually brachycephalic or scaphocephalic with frontal and parietal bossing; micrognathia; microstomia with thin lips; apparently low-set ears; "double chin."

Psychomotor and growth development. Proportionately short stature; intelligence ranges from normal to mental retardation.

Limbs. Occasional syndactyly; winging of the scapulae. See *Comments.*

Other findings. Narrow and highly arched palate; delayed ossification of the craniofacial sutures; microcephaly; cardiac defects; hypogenitalism; cryptorchidism; vertebral anomalies (scoliosis, lordosis, spina bifida); funnel chest; glossoptosis.

Etiology. AR. Heterogeneity? AD?

Comments. Described by Audry in 1893, "though he did not observe the complete syndrome" [Gorlin and Pindborg, 1964; page 432].

The limbs and nails are generally described as normal, but Tridon and Thiriet [1966] referred to some cases with limb anomalies (syndactyly, ectrodactyly, aplasia of the fibula, and homolateral tarsal and metatarsal anomalies), one of which also presented "une absence . . . des ongles" (page 47). If the patient with anonychia really has Hallermann-Streiff's syndrome, this information would add a new element to the large list of "occasional findings," moving this syndrome to the 1-2-3 subgroup. Nevertheless, the frequency of nail defect among Hallermann-Streiff's patients would be very low.

The large majority of the cases are isolated and do not reproduce. Only two patients had children: two normal ones in one instance and an affected in the other (a father-daughter transmission); in the last situation, the marriage involved distant consanguinity [Ref. in Steele and Bass, 1970]. A possible instance of normal but consanguineous parents is reported by Jorgenson et al [1975]. Two affected in an inbred sibship of three are also reported [Ref. in Jorgenson, 1979; McK 23410]. These data suggest an AR mechanism of inheritance. However, a few authors support the possibility of an AD pattern accepting the isolated ("sporadic") cases as being due to fresh mutation and calling attention to the above mentioned father-daughter transmission. (Note, however, that the parents were distant relatives.) In spite of our thinking that the AR inheritance may apply to all cases, we would not like to discard the possibility of a heterogeneity (AD and AR).

This condition—as are some others—is included here even though it is not characterized by cardinal signs of typically ectodermal origin; it is covered, however, by our clinical definition of ectodermal dysplasia. The hair, teeth, and skin involvement gives this condition a clear position among the ectodermal dysplasia/malformation syndromes.

Other references. Streiff [1950], Falls and Schull [1960], Fraser and Fried-mann [1967], Judge and Chakanovskis [1971], Dinwiddie et al [1978].

4. GORLIN-CHAUDHRY-MOSS' SYNDROME (McK 23350)

Synonyms. Gorlin's syndrome; craniofacial dysostosis, patent ductus arteriosus, hypertrichosis, hypoplasia of labia majora, dental and eye anomalies.

Hair. Hypertrichosis (pronounced amount of coarse scalp and body hair, especially on the arms, legs, and back).

Teeth. Hypodontia; microdontia; some pulp chambers small or missing.

Nails. No data.

Sweat. Normal.

Skin. No data.

Hearing. Mild bilateral conductive loss.

Eyes. Hyperopia; microphthalmia; horizontal nystagmus; corneal ulcers; defective eyelid development.

Face. Characteristic, with " 'dished out' appearance of middle face" [Gorlin et al, 1960, page 779], ectropion of lower lid, and antimongoloid slant.

Psychomotor and growth development. Short stature.

Limbs. No data.

Other findings. Craniofacial dysostosis; patent ductus arteriosus; hypoplasia of labia majora; highly arched palate; mild umbilical hernia.

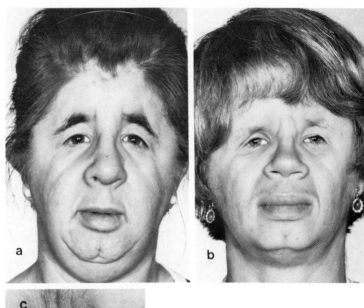

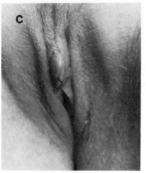

Fig. 8-4. Gorlin-Chaudhry-Moss' syndrome. a,b. Hypertrichosis, antimongoloid slant of palpebral fissures, microphthalmia, and ectropion of lower lip. c. Hypoplasia of labia majora. Courtesy of Dr. Robert J. Gorlin, Minneapolis, MN.

Etiology. AR?

Comments. Gorlin et al [1960] described two sisters from normal nonconsanguineous parents.

Other references. None.

5. MIKAELIAN'S SYNDROME (McK 22480)

Synonyms. Ectodermal dysplasia and neurosensory deafness.

Hair. Coarse and brittle; hypotrichosis of scalp.

Teeth. Carious (patient no. 1).

Nails. Normal.

Sweat. Normal.

Skin. Hyperkeratotic; increased melanin in the basal layer.

Hearing. Bilateral sensorineural loss.

Eyes. Normal.

Face. Coarse features.

Psychomotor and growth development. Slight psychomotor retardation (patient no. 1).

Limbs. Arachnodactyly; contracture of one finger (patient no. 2) and of all fingers (patient no. 1).

Other findings. Kyphoscoliosis (patient no. 2).

Etiology. AR.

Comments. Mikaelian et al [1970] described two affected (one man) in one sibship of ten from first-cousin parents.

Other references. None.

6. GINGIVAL FIBROMATOSIS-SPARSE HAIR-MALPOSITION OF TEETH (McK: not listed)

Synonym. Hair defect-malposition of teeth-gingival fibromatosis.

Hair. Excessively thick in childhood; begins to thin out during early teens; sparse later.

Teeth. Malpositioned and malformed; serrated incisors.

Nails. No data.

Sweat. No data.

Skin. No data.

Hearing. Normal.

Eyes. Alternating strabismus; rotating nystagmus; myopia.

Face. Coarse appearance; protruding lips (secondary to gingival fibromatosis); prognathic mandible; broad and flat nasal alae.

Psychomotor and growth development. Abnormal EEG; low IQ.

Limbs. Large hands; broad and relatively short feet.

Other findings. Highly arched palate.

Etiology. AR.

Comments. Jorgenson [1971] described one patient; she was the daughter of normal remotely consanguineous parents.

Other references. None.

7. HYPERTRICHOSIS AND DENTAL DEFECTS (McK: not listed).

Synonyms. See *Comments.*

Hair. Generalized hypertrichosis (except on palms, soles, and mucous membranes).

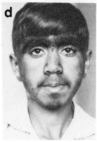

Fig. 8-5. Hypertrichosis and dental defects. a. A 16-year-old boy. b. Father and son. c,d. Mother and son. a,b, Courtesy of Dr. Willy-René Felgenhauer, Neuchâtel, Switzerland; c,d, Freire-Maia et al, 1976.

Teeth. Occasional persistence of deciduous teeth, delayed eruption, hypodontia, anodontia, and supernumerary teeth.

Nails. Normal.

Sweat. Normal.

Skin. Normal.

Hearing. Normal.

Eyes. Normal. See *Comments.*

Face. Normal.

Psychomotor and growth development. Normal. See *Comments.*

Limbs. Normal.

Other findings. See *Comments.*

Etiology. AD. See *Comments.*

Comments. Danforth [1925] and Felgenhauer [1969] referred many cases (mother and daughter; grandfather, daughter, and son; father and son; sporadics) who lived in the past century.

Freire-Maia et al [1976] described a Brazilian mother and son. She had hypertrichosis on scalp, face, axillae, and pubic areas. The boy also had dental abnormalities (supernumerary teeth, persistence of deciduous teeth, and delayed eruption of permanent teeth), growth retardation, low IQ, photophobia, hypotension, and delayed sexual maturation. His low IQ is probably cultural. The reason for the photophobia could not be elucidated since ophthalmological examination was not available. His hypotension, shortness of stature, and delayed sexual maturation could be due to adrenal hypoplasia not related to the hypertrichosis gene, but no endocrinological investigation had been possible in the region where the patients lived.

The above description refers only to a condition characterized by hypertrichosis and dental defects. It should not obscure the fact that hypertrichosis sensu lato is actually a group of conditions rather than a single entity. Hypertrichosis may be congenital or not, localized or generalized, isolated or associated with other traits (metabolic errors, dysplasias and/or malformations). Sometimes it is part of well-delineated ectodermal dysplasias [as defined by Freire-Maia, 1971, 1977a] such as gingival fibromatosis and hypertrichosis (see below); Coffin-Siris' syndrome; Schinzel-Giedion's syndrome; hairy elbows dysplasia; hypomelanosis of Ito; syndrome of accelerated skeletal maturation, failure to thrive, and peculiar face; and congenital lymphedema, hypoparathyroidism, nephropathy, prolapsing mitral valve, and brachytelephalangy. Thus, the whole group of hypertrichotic conditions (ectodermal dysplasias or not) seems to be larger than that which comprises the eight conditions mentioned in this book.

It would be hazardous for us to try to further divide the whole group into more than eight entities because most of the cases referred to in the literature are poorly described, phenotypic overlap between apparently different conditions is plausible, the existence of clinical variants of the same condition is possible, and many situations are not covered by our definition of ectodermal dysplasias. However, we would like to call attention to the fact that this "splitting" task will probably prove to be as important here as it was in several other situations.

The condition that we here call "hypertrichosis and dental defects" has been described several times, but it is doubtful whether all the synonyms applied to hypertrichosis sensu lato would really apply to it. Some of these synonyms are congenital hypertrichosis lanuginosa syndrome, hypertrichosis universalis (McK 14570), hypertrichosis universalis congenita, and edentate hypertrichosis.

Other references. Berres and Nitschke [1968], Cat et al [1971], Suskind and Esterly [1971]. See also McK 14570.

8. GINGIVAL FIBROMATOSIS AND HYPERTRICHOSIS (McK 13540)

Synonyms. Hirsutism and gingival enlargement; excessive hair and gingival enlargement; hypertrichosis with hereditary gingival hyperplasia; syndrome of gingival hyperplasia, hirsutism, and convulsions.

Hair. Generalized hypertrichosis; black and coarse hair in adulthood; bushy eyebrows.

Teeth. Delayed eruption; occasional macrodontia and premature exfoliation.

Nails. Normal.

Sweat. Normal.

Skin. Occasional pigmented nevi and hyperelasticity.

Hearing. No data.

Eyes. No data.

Face. Occasional large ears, peculiar nose, and coarse features.

Psychomotor and growth development. Mental retardation; epilepsy.

Limbs. No data.

Other findings. Gingival fibromatosis; occasional hypoplastic breasts.

Etiology. AD. Heterogeneity? AR? See *Comments.*

Comments. There is not necessarily a relationship between the age of development of the gingival fibromatosis and hypertrichosis. The latter may be present at birth but often appears at puberty, after the development of the gingival enlargement.

According to Witkop [1979], "a recessive form may exist as several families have had sibs but not parents affected" (page 463). If no inbred cases exist, the presence of more than one affected in the same sibship could be due to gonial mutation. However, the possibility of an AR mechanism of inheritance is also considered.

Patients with this condition have a high risk of epilepsy and mental retardation (see above) in contrast to the patients with gingival fibromatosis without hypertrichosis. About one-half (22/52) of the reported cases have had either mental retardation or epilepsy or both [Witkop, 1979].

Oligophrenia and epilepsy seem to be more frequent among nonfamilial cases [Witkop, 1971]. If this is not due to chance, it would add another source of heterogeneity into the group (see the above-mentioned possibility of a recessive form).

Other references. Snyder [1965], Anderson et al [1969], Winter and Simpkiss [1974].

9. PILI TORTI AND ENAMEL HYPOPLASIA (McK 26190)

Synonyms. Twisted hairs; Strandberg-Ronchese's dysplasia; pili torti hereditaria (Ronchese type).

Hair. Pili torti; monilethrix; thin, dry, and blond; when only a part of the scalp is affected, these hairs are of lighter color than the adjacent ones;

bald areas; thin and twisted eyebrow hairs (their outer thirds may be almost entirely absent).

Teeth. Enamel hypoplasia; widely spaced teeth; irregular in shape; serrated biting edges; hypodontia; anodontia; delayed first dentition.

Nails. Normal.

Sweat. Normal.

Skin. Generalized keratosis pilaris.

Hearing. Normal.

Eyes. Chronic blepharitis (?).

Face. Normal.

Psychomotor and growth development. Normal.

Limbs. Normal.

Other findings. No data.

Etiology. AD.

Comments. According to Kurwa and Abdel-Aziz [1973], congenital pili torti may be found in three conditions: Ronchese type (AR), Beare type (AD), and Menkes type (X-linked recessive). However, Ronchese [1932] described only a sporadic case whose parents were normal and nonconsanguineous.

Structural changes are seen in many conditions described in this book, as follows: pili torti—Rapp-Hodgkin's syndrome, Šalamon's syndrome, arthrogryposis and ectodermal dysplasia, ectodermal dysplasia with syndactyly, trichoondotoonychodysplasia with pili torti, pili torti and enamel hypoplasia, and pili torti and onychodysplasia; trichorrhexis nodosa—ichthyosiform erythroderma-deafness-keratitis, and Šalamon's syndrome; trichorrhexis fissurata—Lenz-Passarge's dysplasia; trichorrhexis—onychotrichodysplasia with neutropenia.

Bjørnstad's dysplasia (referred to as Bjørnstad's syndrome; Robinson and Johnston [1967]; Voigtländer [1979]) is characterized by the presence of pili torti and sensorineural deafness. It is, therefore, an ectodermal dysplasia of Group B.

We prefer to consider the conditions described by Strandberg [1922], Ronchese [1932], and Appel and Messina [1942] as a single condition with the name "pili torti and enamel hypoplasia." As mentioned, Ronchese's patient was a sporadic case, but this may represent fresh mutation. The classification of the condition he described together with those of Strandberg [1922] and Appel and Messina [1942] due to AD etiology is mostly based on clinical grounds.

The Beare type belongs to the 1–3 subgroup and will appear under the designation of "pili torti and onychodysplasia."

Menkes et al [1962] described an X-linked recessive condition in five men with pili torti, monilethrix and trichorrhexis nodosa plus severe neurological and mental deterioration, seizures, marked impairment of weight gain, and death in infancy. In one patient at the age of 18 months, "teeth never erupted, although roentgenograms of the mandible showed them to be present" (page 765). Since tooth involvement is represented only by delayed eruption of deciduous teeth in one patient, we prefer to mention this condition here without including it as a separate entity in this (1–2) subgroup; it may belong to the Group B of ectodermal dysplasias, being a member of the 1–5 subgroup.

For an analysis of pili torti (congenital or acquired) as a simple condition, see Kurwa and Abdel-Aziz [1973].

Gorlin and Pindborg [1964] speak of a condition designated anodontia and monilethrix; under this designation they classify together the conditions described by Strandberg [1922] and Rousset [1952]. Rousset's condition (with ungual involvement) seems to be Fischer-Jacobsen-Clouston's syndrome.

Other references. None.

10. WALBAUM-DEHAENE-SCHLEMMER'S SYNDROME (McK: not listed)

Synonyms. None.

Hair. Thin, blond, and sparse in the first years; alopecia later. Hypotrichosis of body.

Teeth. Hypodontia; supernumerary teeth; microdontia; malposition; delayed eruption of permanent dentition; crown dysplasia or root hypoplasia in some teeth.

Nails. Normal.

Sweat. Normal.

Skin. Abnormal dermatoglyphics. See *Comments.*

Hearing. No data.

Eyes. No data.

Face. Swollen, with flat nasal bridge and enlarged tip of nose.

Psychomotor and growth development. Growth retardation.

Limbs. No data.

Other findings. Mild gingival hypertrophy.

Etiology. AR.

Comments. Walbaum et al [1971] described one sibship (F = 1/16) with eight members (six women), two of whom (both women) were affected. A

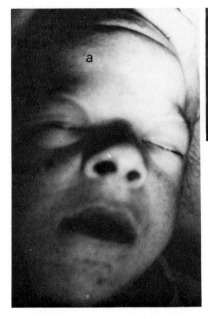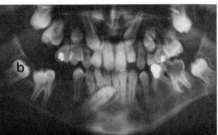

Fig. 8.6. Walbaum-Dehaene-Schlemmer's syndrome. a. Hypotrichosis of scalp (the patient wears a wig), eyebrows, and lashes; swollen face. b. Extensive dental abnormalities. Courtesy of Dr. Roland Walbaum, Lille, France.

schematic drawing of palmar and finger dermatoglyphics was presented without analysis. Some disturbances can be seen in the palmar impressions: absence of loops, presence of accessory triradii, etc.

Other references. None.

11. OCULOOSTEOCUTANEOUS SYNDROME (McK 25790)

Synonyms. Brachymetapody, anodontia, hypotrichosis and albinoid trait; Tuomaala-Haapanen's syndrome.

Hair. Scanty, thin, and blond scalp hair; absent pubic and axillary hair; thin eyelashes.

Teeth. Anodontia.

Nails. Normal.

Sweat. Normal.

Skin. Blond, with little elasticity; hypoplastic nipples and unpigmented areolae; palmar hyperkeratosis.

Hearing. No data.

Eyes. Convergent strabismus; distichiasis; horizontal nystagmus; foveal hypoplasia; lenticular and corneal opacity; severe myopia.

Face. Broad base of the nose; low-set and hypoplastic ear lobes; antimongoloid palpebral fissures; flaccid tarsus of upper lid; prominent mandible; hypoplastic maxilla.

Psychomotor and growth development. Short stature.

Limbs. Lower micromelia; brachychiry. With the exception of the big toes (which are excessively large and "clubbed"), all toes, especially the third, fourth, and fifth, are short; some of them are malpositioned.

Other findings. Short skull (high cephalic index); small breasts; underdeveloped external genitalia.

Etiology. AR?

Comments. Tuomaala and Haapanen [1968] described an apparently noninbred sibship of six (four women), three of whom (two women) were affected.

Other references. None.

12. AGAMMAGLOBULINEMIA-THYMIC DYSPLASIA-ECTODERMAL DYSPLASIA (McK: not listed)

Synonyms. None

Hair. Alopecia and absence of eyebrows and lashes.

Teeth. Natal; hypoplastic; enamel defects. See *Comments.*

Nails. No data.

Sweat. No data.

Skin. In the neonatal period, several bullous lesions on face, hands, and feet; thin, dry, and scaly; ichthyosiform erythroderma involving the entire body; atrophic epidermis.

Hearing. No data.

Eyes. No data.

Face. Normal.

Psychomotor and growth development. No data.

Limbs. Normal.

Other findings. Hypoplastic and histologically abnormal thymus; agammaglobulinemia; low IgG and IgA levels.

Etiology. Unknown.

Comments. Feigin et al [1971] described a noninbred girl who died at 7 months. At 5 months, she had six teeth that "showed enamel defects and appeared hypoplastic" (page 144). Her father had a low IgM level.

For agammaglobulinemia associated with short-limb dwarfism and ectodermal dysplasia, see Chapter 9.

Other references. None.

13. JOHANSON-BLIZZARD'S SYNDROME (McK 31048)

Synonym. Syndrome of congenital aplasia of the alae nasi, deafness, hypothyroidism, dwarfism, absent permanent teeth, and malabsorption.

Hair. Sparse, dry, and fine or coarse; marked frontal upsweep.

Teeth. Oligodontia of both dentitions; peg-shaped teeth.

Nails. No data.

Sweat. Normal.

Skin. Pale and smooth; café-au-lait spots on the lower limbs and abdomen; patches of vitiligo on the lower back and abdomen; midline scalp defects (aplasia cutis congenita); tiny nipples with almost no areolae; transverse palmar creases.

Hearing. Congenital sensorineural deafness.

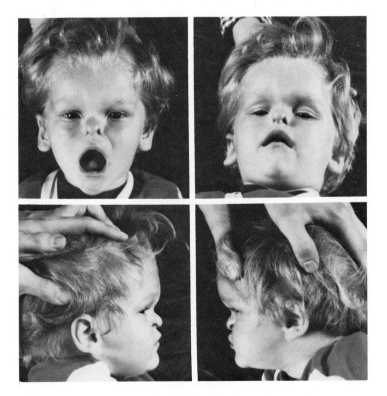

Fig. 8-7. Johanson-Blizzard's syndrome. Abnormal auricles, absent nasal alae, small nose with beaklike appearance, upsweep of frontal hair, hypoplastic midface and fine hair. Courtesy of Dr. Donald W. Day, Chicago, IL.

Eyes. Aplasia of the inferior puncta; strabismus.

Face. Aplastic alae nasi that give the nose a beaklike appearance.

Psychomotor and growth development. Severely retarded; hypotonia; abnormal EEG; occasional akinetic seizures.

Limbs. No data.

Other findings. Microcephaly; hypothyroidism; pancreatic dysfunction; imperforate anus; genito-urinary defects (rectovaginal fistula; double or septate vagina; single urogenital orifice; clitoromegaly; immature ovaries; micropenis); failure to thrive/edema; malabsorption; epiphyseal dysgenesis; nasolacrimo-cutaneous fistulae; highly arched palate; delayed bone age; hyperextensibility.

Etiology. AR.

Comments. Johanson and Blizzard [1971] described three women from normal nonconsanguineous parents, belonging to different families. Day and Israel [1978] reviewed the literature and added two new cases. Consanguinity was found for about half of the cases.

Other references. Mardini et al [1978], Daentl et al [1979], Baraitser and Hodgson [1982].

14. TRICHODENTAL DYSPLASIA (McK: not listed)

Synonym. Tricho-dental syndrome.

Hair. Straight, fine, and lusterless appearance probably due to lack of pigmentation; relatively thin shafts with a slight beading effect due to variations in the shaft contour; abnormal cuticle scale pattern and missing scale; scanty or absent distal eyebrows and sparse eyelashes.

Teeth. Hypodontia; peg-shaped teeth; retained deciduous teeth.

Nails. Normal.

Sweat. No data.

Skin. No data.

Hearing. No data.

Eyes. No data.

Face. No data.

Psychomotor and growth development. No data.

Limbs. No data.

Other findings. No data.

Etiology. AD.

Comments. Salinas and Spector [1979, 1980] described two families with affected men and women over four and three generations, respectively. No transmission from father to son is mentioned although a bald man (not

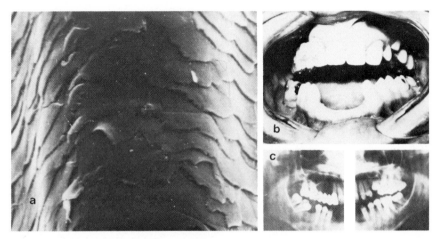

Fig. 8-8. Trichodental dysplasia. a. Pinched appearance as revealed by scanning electron microscopy. b,c. Hypodontia and peg-shaped teeth; note a retained upper left primary molar. Courtesy of Dr. Carlos F. Salinas, Charleston, SC.

mentioned as having tricho-odontic anomalies) had two affected sons from two marriages.

The categories above for which there is no data may be interpreted as normal, though normality was not referred to in the papers [Salinas, 1982; pers. communication].

Other references. None.

15. ACRORENAL-ECTODERMAL DYSPLASIA-LIPOATROPHIC DIABETES (AREDYLD) SYNDROME (McK: not listed)

Synonyms. None.

Hair. Scalp hypotrichosis (fine, slow-growing hair); scant axillary and pubic hair; normal eyebrows and lashes.

Teeth. Two natal and four deciduous teeth with enamel dysplasia (all of them lost in infancy); absence of permanent teeth buds; anodontia at age 11.

Nails. Normal. See *Comments.*

Sweat. Normal.

Skin. Right hypoplastic and left hypopigmented areolae (both with somewhat diffuse limits); absence of DIP extension and flexion creases.

Hearing. Normal.

Eyes. Normal.

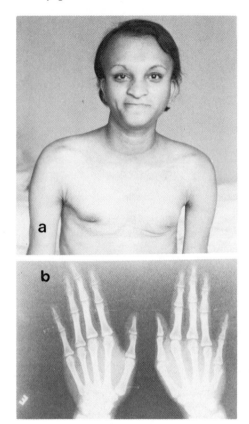

Fig. 8-9.   Acrorenal-ectodermal dysplasia-lipoatrophic diabetes syndrome. a. Scalp hypotrichosis, peculiar face, aplastic and hypoplastic mammary glands, and hypoplastic nipples.
b. Fourth and fifth right metacarpal bones shorter than normal. Unpublished photos.

Face. Prominent forehead and bridge of nose; slight mongoloid slant of
palpebral fissures; short nasal septum with flat tip of nose; short upper lip;
relatively flat philtrum; prominent chin with mandibular prognathism;
posteriorly angulated auricles with broad intertragal incisure; hypoplastic
tragus and small groove at antitragus.

Psychomotor and growth development. Short stature; difficulty in grasping
with left hand.

Limbs. Shortness of the right fourth and fifth metacarpal bones; rarefac-
tion of the right tibia and of the bones of the right foot; medial bending of
the long axis of the right tibia; mild right pes varus; bilateral pes planus;

left second toe larger than the corresponding hallux; shortening of ischium and hypoplastic ilium.

Other findings. Lipoatrophic diabetes; hypoplastic labia minora; hypertrophy of clitoris; small vaginal introitus; absence of right and hypoplasia of left mammary gland; lumbar scoliosis; hyperostosis of cranial vault; cranial dysostosis; prominent subcutaneous leg veins; hypoplasia of the middle right major renal calyx and hypotonia of the right ureter. See *Comments*.

Etiology. AR.

Comments. Pinheiro et al [1983b] described a 22-year-old girl with the above clinical picture. A similar picture of ectodermal dysplasia was also present in one of her sisters who died at 1 year of age. The parents were normal and second cousins (F = 1/64).

Prognathism, hypertrophy of clitoris, and prominent subcutaneous leg veins seem to be consequences of lipoatrophic diabetes.

Nails are certainly normal, "but the fifth fingernail is always angling downward" [Pinheiro et al, 1983b; page 32].

This syndrome has some similarities with Werner's syndrome (McK 27770) but the differences are sufficiently large to permit accepting it as a different condition.

Other reference. Pinheiro et al [1982, 1983a].

16. ALOPECIA-ANOSMIA-DEAFNESS-HYPOGONADISM (McK: not listed)

Synonyms. None.

Hair. Absent or sparse scalp hair, eyebrows, and lashes, axillary and pubic hair.

Teeth. Carious, leading to extensive premature loss.

Nails. Normal.

Sweat. Normal.

Skin. Normal.

Hearing. Conductive loss. See *Comments*.

Eyes. Normal.

Face. Unilateral palsy; retro/micrognathia; auricle abnormalities (see *Comments*).

Psychomotor and growth development. Anosmia or hyposmia; mental retardation; speech impairment; hypotonia.

Limbs. Normal.

Other findings. Hypogonadism; occasional congenital heart defects; cleft palate; choanal stenosis.

Etiology. AD.

Comments. Johnson et al [1983] described ten men and six women in a three-generation kindred.

Ear defects included atresia of auditory canal, microtia and asymmetric/prominent ears.

Other reference. Johnson et al [1982].

9. Subgroup 1-3

1. HAIRY ELBOWS DYSPLASIA (McK 13960)

Synonyms. Familial hypertrichosis cubiti; hypertrichosis cubiti; hairy elbows.

Hair. Hypertrichosis (the hair is long, dark, and coarse) of the elbow region, involving the lower third of the arm and the upper third of the forearm; it is more pronounced in infancy and then regresses in later childhood. See *Comments.*

Teeth. Normal.

Nails. Short (but not dysplastic) fingernails. See *Comments.*

Sweat. Normal.

Skin. Normal.

Hearing. No data.

Eyes. No data.

Face. Normal.

Psychomotor and growth development. Short stature. See *Comments.*

Limbs. No data.

Other findings. No data.

Etiology. AR. See *Comments.*

Comments. Beighton [1970] described a boy and a girl from a highly inbred Old Order Amish sibship of five.

The presence of hairiness in the father and other members of their family seems not to be related to this dysplasia. The children were also short of stature, but this trait may also be unrelated to their hypertrichosis.

Andreev and Stransky [1979] described a boy in a sibship of two from normal nonconsanguineous parents. They did not mention nail involvement.

The hypertrichosis that was noted soon after birth increased in length and thickened in texture during infancy. At the age of 5 it reached a maximum; then the condition slowly regressed.

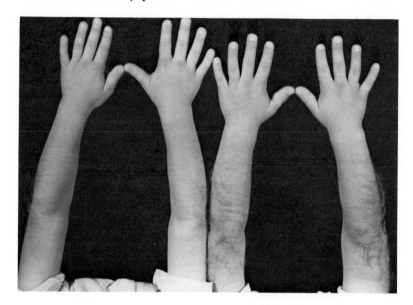

Fig. 9-1.   Hairy elbows dysplasia. Hypertrichosis of the elbow regions and shortness of nails. Courtesy of Dr. Peter Beighton, Cape Town, South Africa.

Other references. None.

2. PALMOPLANTAR HYPERKERATOSIS AND ALOPECIA (McK 10410)

Synonyms. Alopecia congenita with keratosis palmo-plantaris.

Hair. Hypotrichosis to alopecia; absence of eyebrows and lashes; hypotrichosis of axillae and pubic regions.

Teeth. Normal.

Nails. Short and dystrophic with onycholysis.

Sweat. Normal.

Skin. Palmoplantar hyperkeratosis.

Hearing. No data.

Eyes. No data.

Face. Normal.

Psychomotor and growth development. Normal.

Limbs. No data.

Other findings. No data.

Etiology. AD. Heterogeneity? AR? See *Comments.*

Comments. Stevanović [1959] described five patients (three men) over three generations. He suggested the hypothesis of an AD gene with incomplete penetrance and variable expressivity. Two of the five patients were reported to have the diffuse keratoderma of the Unna-Thost type (early onset), and another one, the striated Brünauer-Fuhs type (later onset).

Podoswa-Martinez et al [1973] described a condition clinically similar to that reported by Stevanović [1959], but claimed an AR mode of inheritance for it. The parents of the sibship of five (three men and two women; only a normal woman) were nonconsanguineous. In our genetic analysis (Chapter 16) we will accept heterogeneity as possible (AD and AR?).

According to Gorlin et al [1964], the condition described by Stevanović [1959] is the same as reported by Unna, in 1883, and by Spannlang, in 1927 (Spannlang-Tappeiner's syndrome). The signs of this AD condition are partial or complete alopecia, hyperhidrosis, tongue-shaped corneal opacities, and palmoplantar keratosis. Also according to Gorlin et al [1964], the patients described by Brünauer, in 1925, had another condition (Brünauer's syndrome), with mental retardation, defective enamel, hyperhidrosis, and palmoplantar keratosis.

Other references. None.

3. CURLY HAIR-ANKYLOBLEPHARON-NAIL DYSPLASIA (CHANDS) (McK 21435)

Synonyms. None.
Hair. Curly (7/7).
Teeth. Normal.
Nails. Hypoplastic finger- and toenails (7/7).
Sweat. Normal.
Skin. No data.
Hearing. Normal.
Eyes. Fused eyelids at birth (ankyloblepharon) (4/7).
Face. Normal.
Psychomotor and growth development. Normal.
Limbs. No data.
Other findings. No data.
Etiology. AR.
Comments. Baughman [1971] described seven patients (three men) with this condition and suggested an AD inheritance for it. A better analysis of the family led Toriello et al [1979] to postulate an AR etiology. The patients belong to two inbred sibships with a total of 17 persons; two affected

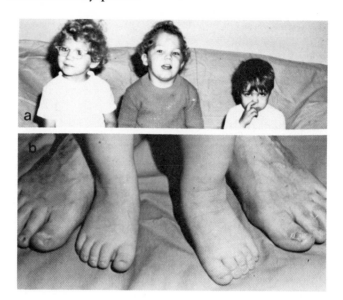

Fig. 9-2.   Curly hair-ankyloblepharon-nail dysplasia. a. Curly hair (the third child is normal). b. Onychodysplasia in father and daughter. Courtesy of Dr. Fred A. Baughman, Jr., San Diego, California.

persons had a total of 11 normal children; the only affected with affected children married a first and third cousin.

Other references. None.

4. ONYCHOTRICHODYSPLASIA WITH NEUTROPENIA (McK 25836)

Synonym. Onychotrichodysplasia with chronic neutropenia.

Hair. Dry, fine, lusterless, short, curly, and sparse on scalp, eyebrows, and lashes; trichorrhexis; absent at axillae and pubic areas at puberty.

Teeth. Normal.

Nails. Hypoplastic finger- and toenails; koilonychia; onychorrhexis.

Sweat. Normal.

Skin. Thin and wrinkled on the palms and soles; pustules on the palms.

Hearing. Normal.

Eyes. Chronic irritative conjunctivitis.

Face. Normal.

Psychomotor and growth development. Delayed psychomotor development; mild generalized hypotonia.

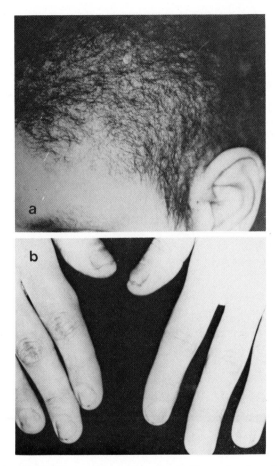

Fig. 9-3.   Onychotrichodysplasia with neutropenia. a. Short, dry, lusterless, curly, and sparse scalp hair and eyebrows. b. Dystrophic nails. Courtesy of Dr. Alejandro Hernández, Guadalajara, Jalisco, Mexico.

Limbs. Normal.

Other findings. Chronic neutropenia; recurrent infections (mainly tonsillitis, sinusitis, and otitis).

Etiology. AR.

Comments. Cantú et al [1975] described a boy in a sibship of three (all men), from half first-cousin parents (F = 1/32). Hernandez et al [1979] reported three sisters who also had consanguineous parents (F = 5/64).

Other references. None.

5. PILI TORTI AND ONYCHODYSPLASIA (McK: not listed)

Synonym. Pili torti (Beare type).

Hair. Normal in infancy followed by hypotrichosis on the scalp, and in the axillary and pubic areas; coarse (5/5); pili torti.

Teeth. Normal.

Nails. Short, fragile, and brittle ($\geq 4/5$).

Sweat. Normal. See *Comments.*

Skin. Normal (3/5), dry (1/5), or greasy (1/5). Slight atrophy on the top of the scalp (1/5).

Hearing. Normal.

Eyes. Normal.

Face. Normal.

Psychomotor and growth development. Low IQ (4/5); severe mental retardation (1/5); "irresponsible personality" (5/5).

Limbs. Normal.

Other findings. No data.

Etiology. AD.

Comments. Beare [1952] described a man and four of his eight children (one man). One of the patients "complains of excessive sweating of all areas" (page 368). However, no sweat test was applied. See *Comments* in Pili torti and enamel hypoplasia.

Other references. None.

6. AGAMMAGLOBULINEMIA-DWARFISM-ECTODERMAL DYSPLASIA (McK: not listed; see *Comments*)

Synonyms. None.

Hair. At birth, scalp hair, eyebrows, and lashes are abundant, but in the ensuing months all hair falls out and fails to grow back.

Teeth. No data. See *Comments.*

Nails. Small and dystrophic.

Sweat. No data.

Skin. Erythroderma; mild hyperkeratosis; generalized scaliness; ichthyosiform lesions; redundant, especially on the limbs, suggesting cutis laxa. Biopsy showed keratosis, fissuring of keratotic layer, and thickening of granular layer.

Hearing. No data.

Eyes. No data.

Face. No data.

Psychomotor and growth development. Dwarfism.

Limbs. Dyschondroplastic (short-limbed) dwarfism.

Other findings. Lymphopenic agammaglobulinemia; prominent eosinophilia; hypoplastic thymus; microscopic alterations of thymus, spleen, lymph nodes, gastrointestinal tract, bones, etc.

Etiology. AR?

Comments. Gatti et al [1969] described a brother-and-sister pair from normal nonconsanguineous parents; the patients died at the ages of about 5 and 8 months, respectively; no information was provided regarding the presence or absence of tooth buds. The question of dental involvement requires further investigation.

Besides being an ectodermal dysplasia, this condition is also a chondrodysplasia. However, it has distinctive aspects separating it from all other forms of short-limb dwarfisms.

There is a condition of agammaglobulinemia and an achondroplasia-like skeletal disorder, without ectodermal dysplasia [McKusick and Cross, 1966; Davis, 1966; Fulginiti et al, 1967] and another one of agammaglobulinemia, thymic dysplasia and ectodermal dysplasia, without skeletal disorders [Feigin et al, 1971].

This condition is referred to in McK 20090 under the label of achondroplasia and Swiss-type agammaglobulinemia.

Other references. None.

7. TRICHOONYCHODYSPLASIA WITH XERODERMA
(McK: not listed)

Synonyms. None.

Hair. Congenital universal atrichia; later normal.

Teeth. Normal.

Nails. Congenital anonychia; very convex fingernails without eponychium; dystrophic toenails: two with the free margins growing downwards close to the skin and one (that of the fifth left toe) thickened and darkened.

Sweat. Normal.

Skin. Severe xeroderma with permanent and abundant scaling over the entire body. Tendency toward fissures in hands and feet.

Hearing. Normal.

Eyes. Normal.

Face. Normal.

Psychomotor and growth development. Normal.

Limbs. Normal.

Other findings. No data.

Etiology. AR.

Comments. Freire-Maia et al [1982] described one boy from normal nonconsanguineous parents. The sibship also includes an affected girl (died at 3 months) and three normal members (one man). A first and second cousin of the proband (a girl), the offspring of a consanguineous marriage (F = 1/16), is also reported to be more severely affected. The mother of the propositus presents discrete xeroderma.

Other reference. Freire-Maia et al [1984].

8. SKELETAL ANOMALIES-ECTODERMAL DYSPLASIA-GROWTH AND MENTAL RETARDATION (McK: not listed)

Synonyms. None.

Hair. Almost complete absence of body hair; a few curled hairs are present in the parieto-occipital and pubic regions.

Teeth. Normal. See *Comments.*

Nails. Hypoplastic toenails.

Sweat. Normal.

Skin. Dry and hyperkeratotic with rhomboid type of scaling, especially on the lower legs; absence of flexion creases on thumbs; fifth fingers with single flexion creases.

Hearing. Normal.

Eyes. Normal.

Face. Large, prominent nose; upslanting palpebral fissures; short upper lip; large and poorly formed ears.

Psychomotor and growth development. Proportionate, extreme growth failure; severe mental retardation.

Limbs. Narrow and small hands with short and distally tapered fingers; the fifth fingers are particularly short; short midphalanges (especially of fingers 2 and 5); flexion contraction of elbows; hip luxation; small feet, with dorsiflexion of the left hallux; bilateral broad gap between first and second toes; bilateral complete cutaneous syndactyly between fourth and fifth toes; hypoplastic little toes; short metatarsals of big toes. The left leg is short, standing in anteposition and inward rotation; its muscles are hypotrophic. Bone fusions: talus-navicular, metacarpals 4 and 5, lunate-triquetral. Humero-radial ankylosis.

Other findings. Microbrachycephaly; multiple uni- or bilateral fusion of vertebral bodies in the lower thoracic and upper lumbar regions.

Etiology. Unknown.

Comments. Schinzel [1980] described a 17-year-old woman from normal nonconsanguineous parents. She had four normal sibs. This patient was

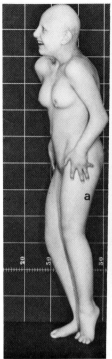
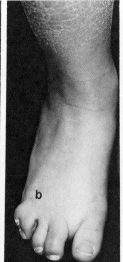
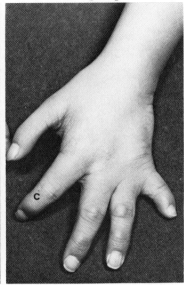

Fig. 9-4. Skeletal anomalies-ectodermal dysplasia-growth and mental retardation. a. Microbrachycephaly, generalized and severe hypotrichosis, large, prominent nose, short left leg, and left hammer toe. b. Hyperkeratosis, broad gap between first and second toe, hypoplastic nails, and cutaneous syndactyly between fourth and fifth toes. c. Narrow hand, short fifth finger with a single flexion crease, and absence of flexion crease at the thumb. Courtesy of Dr. A. Schinzel, Zürich, Switzerland.

compared to those described by Bass et al [1975] and Majewski and Spranger [1976]; although the three present marked similarities, Schinzel [1980] refrained from classifying them as being the same syndrome due to the differences he found between his patient and the other two.

The teeth are mentioned here as normal and they really look so in Figure 1b of Schinzel [1980], who described them as "crowded, but normally shaped and not notably carious" (page 246).

Other references. None.

9. SABINAS BRITTLE HAIR AND MENTAL DEFICIENCY SYNDROME (McK 21139)

Synonyms. Brittle hair and mental deficit; Sabinas brittle hair syndrome.

Hair. Dry, brittle, coarse, wiry in texture, and breaking off easily with combing and brushing; alopecia in areas of pillow contact; decreased cuticular layer and an apparently collapsed cortex; reduced cystine content and increased copper/zinc ratio; decreased lysine, aspartic acid, alanine, leucine, isoleucine, and tyrosine contents; reduced eyebrows and lashes; absence of axillary and pubic hair.

Teeth. See *Comments.*

Nails. Dystrophic (splitting and cracking proximally).

Sweat. Normal.

Skin. Scalp hyperkeratosis on exposed areas.

Hearing. Normal.

Eyes. Different patients present pigmentary retinopathy, unilateral congenital tortuosity of the retina vessels, pale optic discs, and hyperopic astigmatism.

Face. Maxillary hypoplasia. See *Comments.*

Psychomotor and growth development. Mental retardation; deficiency in eye-hand coordination in one patient.

Limbs. Normal.

Other findings. Delayed bone age.

Etiology. AR.

Comments. Arbisser et al [1976] described a brother-and-sister pair from normal nonconsanguineous parents. Howell et al [1981] reexamined both patients and described others in a total of 12 (six women) belonging to five sibships from apparently different families; one of these sibships is inbred (F = 1/16).

No detailed description of the teeth was provided; they were incidentally mentioned as "crowded."

Howell et al [1981] noted that "the brittle, short hair, reduced eyelashes, crowded teeth, and dull appearance created a characteristic facial appearance" (page 959).

The name of this syndrome comes from the name of the small, remote village in Northern Mexico where the patients were born.

Other references. None.

10. SYNDROME OF ACCELERATED SKELETAL MATURATION, FAILURE TO THRIVE, AND PECULIAR FACE (McK: not listed)

Synonyms. Marshall's syndrome II (see *Comments*); syndrome of accelerated skeletal maturation and relative failure to thrive; syndrome of accelerated skeletal maturation in infancy, peculiar facies, and multiple congenital anomalies.

Hair. Hypertrichosis; coarse eyebrows.

Teeth. No data.

Nails. Spoon-shaped.

Sweat. No data.

Skin. Occasionally loose.

Hearing. No data.

Eyes. Shallow orbits; blue sclerae; megacornea; microphthalmia.

Face. Low-set floppy ears; hypertelorism; micrognathia; antimongoloid palpebral slanting; small nose; long philtrum; microstomia; prominent forehead; upturned nares; large ears; proptosis.

Psychomotor and growth development. Both retarded; failure to thrive in terms of height and weight.

Limbs. Abnormal middle phalanges in hands; camptodactyly; clinodactyly; long hands or fingers; long feet; long thin tubular bones; low-set thumbs; prehensile halluces; prominent heels.

Other findings. Accelerated skeletal maturation; choanal atresia; narrow thorax; scoliosis; increased skull density; small facial bones; respiratory problems; feeding difficulties; highly arched palate; hydronephrosis; early death.

Etiology. AR?

Comments. There are two Marshall's syndromes: this one, described by Dr. R.E. Marshall and co-workers [Marshall et al, 1971], and that described by Dr. Don Marshall [Marshall, 1958]. They are quite different. Reference to both under the same eponym should require the designation Marshall's syndrome I for that of Dr. Don Marshall and Marshall's syndrome II for the other.

Marshall's syndrome II and Weaver's syndrome [Weaver et al, 1974] have been accepted as the same condition by some authors. According to Fitch [1980], they are different entities, and a differential diagnosis should consider the following items: Marshall's syndrome II—prominent calvarium, small facial bones, proptosis, flat nasal bridge, upturned nares, failure to thrive, respiratory problems, and early death; Weaver's syndrome— overgrowth, hypertonia, increased bifrontal diameter, and prominent fingerpads with thin, deep-set nails. The apparent growth potential of the children seems to represent an important guide for separating both syndromes [Smith, 1974].

Other reference. Visveshwara et al [1974].

11. CONGENITAL LYMPHEDEMA, HYPOPARATHYROIDISM, NEPHROPATHY, PROLAPSING MITRAL VALVE, AND BRACHYTELEPHALANGY (McK: not listed)

Synonyms. None.

Hair. Hypertrichosis on face and forehead; medial flare of the eyebrows. See *Comments.*

Teeth. No data.

Nails. Short nail beds.

Sweat. No data.

Skin. Thick and dry on lymphedema areas.

Hearing. Normal.

Eyes. Bilateral cataracts.

Face. Mild palpebral ptosis; broad nasal bridge; lateral displacement of the inner canthi.

Psychomotor and growth development. Short stature.

Limbs. Lymphedema of the upper and lower limbs; brachytelephalangy; increased carrying angle.

Other findings. Small and inadequately functioning kidneys with smooth outlines; prolapsing mitral valve (see *Comments*); congenital hypoparathyroidism; probable pulmonary lymphangiectasia.

Etiology. AR? XR? See *Comments*.

Comments. Dahlberg et al [1983] described two brothers in a sibship of four (three men) from normal nonconsanguineous parents. Since the affected have a half maternal brother with slight hypertrichosis of the forehead and a medial flare of the eyebrows, an X-linked inheritance is plausible. The paternal grandfather has a prolapsing mitral valve. The medial flare of eyebrows is considered as a "normal familial variant" (page 103).

In spite of the fact that the cardinal signs are not ectodermal, this condition is mentioned here because it is covered by our definition of ectodermal dysplasia of Group A.

Other references. None.

10. Subgroup 1-4

1. ECTODERMAL DYSPLASIA OF THE HEAD (McK 13650)

Synonyms. Congenital ectodermal dysplasia of the face; focal facial dermal dysplasia; hereditary symmetrical aplastic nevi of temples; facial ectodermal dysplasia.

Hair. Alopecia areata (over the lesions); generally sparse eyebrows (in their lateral thirds) and lashes or multiple rows of lashes on the upper lids; normal to absent lashes on lower lids.

Teeth. Normal.

Nails. Normal.

Sweat. Localized hypohidrosis (scarce-to-absent sweat glands in the focal lesions).

Skin. Somewhat roundish focal temporal lesions that have a smooth or wrinkled surface and may be hyperpigmented, and look like forceps marks. Occasional multiple vertical linear depressions on the lower forehead ("guttate lesions"). Absence of sebaceous glands in the temporal lesions.

Hearing. Normal.

Eyes. Chronic bilateral blepharitis in a few cases; exotropia in one case.

Face. Leonine appearance. Wrinkles periorbitally, wide nasal bridge, fleshy nose with the tip bent down, nose and chin rubberylike when palpated, and bilateral epicanthal fold. See *Comments*.

Psychomotor and growth development. Normal.

Limbs. Normal.

Other findings. No data.

Etiology. AD. Heterogeneity? AR?

Comments. Brauer [1929] described 38 patients and traced the condition through five generations of a family in which about 155 persons were said to have been affected. Setleis et al [1963] described five children (one of them a boy) from three noninbred sibships of 19, all of Puerto Rican origin, and suggested an AR inheritance. Jensen [1971] described two families (one with 20 affected individuals over five generations and the other with three affected children in a noninbred sibship of seven). In both of Jensen's families the clinical signs are similar to those mentioned by Brauer [1929], but different from the syndromic picture described by Setleis et al [1963], who are the only authors to mention a characteristic face (see *Face*).

Summing up the above information on the six families, we have the following groupings: 1) Normal face: AD [Brauer, 1929]; AD [Jensen, 1971], family; AR? [Jensen, 1971], family M, segregation ratio 3/7. 2) Characteristic face: AR? [Setleis et al, 1963]. First family, segregation ratio 2/2; second family, segregation ratio 2/10; third family, segregation ratio 1/7. Total (uncorrected) segregation ratio 5/19 = 0.26.

We concur with Jensen [1971] that the term congenital ectodermal dysplasia of the face "should be regarded as a collective description covering possibly more than one inherited condition" (page 416). Our impression is that we have here at least two different syndromes: the AD Brauer-Jensen form and the possibly AR Setleis form. The possibly AR Jensen type may represent either another cryptic syndrome in comparison with the dominant one or the same dominant syndrome (gonadal mutation in a gametocyte precursor on one of the parents?). Anyway, more data are needed to clarify the problem of clinical delineation of the above forms. In our final analysis, we shall consider two different syndromes: an AD one (without characteristic face) and a possibly AR one (AR?) with a characteristic face.

Other references. None.

2. TRICHOFACIOHYPOHIDROTIC SYNDROME
(McK: not listed)

Synonym. Hypohidrosis with sparse hair and short stature.
Hair. Medium texture, sparse, brittle, and hypochromic.
Teeth. Normal.
Nails. Normal.
Sweat. Hypohidrosis (normal number and size of palmar sweat pores).
Skin. No data.
Hearing. No data.
Eyes. Normal.
Face. Unusual, with wide and deeply depressed base of nose, anteverted nostrils, long philtrum, malar hypoplasia, and hypertelorism.
Psychomotor and growth development. Proportionately short stature.

Limbs. No data.

Other findings. Mild upper respiratory problems.

Etiology. XR?

Comments. Antley et al [1976] described a 10-year-old noninbred boy and mentioned a deceased maternal great-uncle seen only in photographs; according to the family, he bore a striking resemblance to the proband.

Other references. None.

11. Subgroup 2-3

1. TRIPHALANGEAL THUMBS-ONYCHODYSTROPHY-DEAFNESS (McK 22050)

Synonyms. Triphalangeal thumbs-hypoplastic distal phalanges-onychodystrophy; deaf-mutism and onychodystrophy; deafness-onychoosteodystrophy-mental retardation; DOOR syndrome.

Hair. Normal.

Teeth. Hypoplastic and discolored; irregular placement.

Nails. Hypoplastic and dystrophic finger- and toenails; anonychia.

Sweat. Normal.

Skin. Dermatoglyphic abnormalities (simple arch patterns practically on all fingers and toes).

Hearing. Sensorineural deafness.

Eyes. Normal.

Face. Apparently low-set ears.

Psychomotor and growth development. Seizures and mental retardation in the recessive form. (See *Comments*).

Limbs. Triphalangy of both thumbs and halluces; hypoplasia or aplasia of terminal phalanges of fingers and toes; occasional clinodactyly and camptodactyly.

Other findings. No data.

Etiology. Heterogeneity. AD and AR.

Comments. Pinsky [1975, 1977] has concluded that the two mother-son pairs described by Goodman et al [1969], and Moghadam and Statten [1972], have an AD syndrome, and that the patients described by Walbaum et al [1970], Qazi and Smithwick [1970], and Cantwell [1975], have a similar condition but with an AR pattern of inheritance.

Sánchez et al [1981] described two daughters in a sibship of four from normal consanguineous parents (F = 1/64).

As mentioned above, seizures and mental retardation would be seen only in the recessive form.

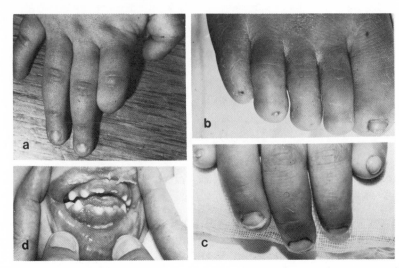

Fig. 11-1. Triphalangeal thumbs-onychodystrophy-deafness. a. Triphalangeal thumbs (not clearly seen), onychodystrophy, anonychia, absence of terminal phalanx. b,c. Anonychia and onychodystrophy. d. Widely spaced and poorly formed teeth. a, courtesy of Dr. Roland Walbaum, Lille, France; b–d, courtesy of Dr. Qutub H. Qazi, New York, NY.

Feinmesser and Zelig [1961] described two inbred patients with congenital deafness associated with onychodystrophy (with normal teeth and fingers). This condition, accepted by several authors as being the same as described here, probably is a different syndrome. According to this view (accepted by Sánchez et al [1981]), we would have here three different conditions, as follows: 1) an AD condition [Goodman et al, 1969, and Moghadam and Statten, 1972]. 2) two AR conditions: a. Walbaum et al [1970]; Qazi and Smithwick [1970]; Cantwell [1975]; Sánchez et al [1981]. b. Feinmesser and Zelig [1961]. This condition does not seem to be an ectodermal dysplasia of group A.
Other references. None.

2. ECTODERMAL DEFECT WITH SKELETAL ABNORMALITIES (McK: not listed)

Synonyms. None.
Hair. Scalp hair is slightly coarse; very sparse axillary hair.
Teeth. Hypodontia (large numbers of unerupted teeth); hypoplastic teeth.
Nails. Finger- and toenails poorly developed and foreshortened.
Sweat. Normal.

Skin. Thin, fine, and dry; fine light granular pigmentation; translucent appearance (with superficial veins showing through the back); rudimentary nipples.

Hearing. No data.

Eyes. No data.

Face. Striking appearance. The central portion is relatively underdeveloped; the cheeks, upper jaw, and nose are sunken, with the "inverted dish-shaped deformity" [Wallace, 1958; page 19] and somewhat prominent eyes.

Psychomotor and growth development. Low intelligence.

Limbs. Short metacarpals; some absorption of the terminal tufts of the distal phalanges; flexion anomalies of hands and feet.

Other findings. Absence of breasts; narrow and highly arched palate.

Etiology. Unknown.

Comments. Wallace [1958] described a woman (aged 22) from normal nonconsanguineous parents. Her 11 siblings (six sisters) are all normal.

The name of this condition is exactly that given by the author in the title of his paper; it is a poor designation, since many syndromes combine ectodermal dysplasia with skeletal defects.

Other references. None.

3. ROBINSON'S SYNDROME (McK 12448)

Synonyms. Deafness, ectodermal dysplasia, polydactylism and syndactylism; ectodermal dysplasia and deafness; peg-shaped teeth, partial anodontia, hypoplastic and dystrophic nails, poly- or syndactyly; deafness and onychodystrophy, dominant form.

Hair. Normal.

Teeth. Hypodontia; coniform teeth; persistence of deciduous teeth; delayed eruption of both dentitions.

Nails. Small and dystrophic with furrows and cracks.

Sweat. Normal. Elevation of sodium and chloride concentrations.

Skin. No data.

Hearing. Sensorineural deafness.

Eyes. No data.

Face. No data.

Psychomotor and growth development. No data.

Limbs. Occasional partial syndactyly and polydactyly.

Other findings. No data.

Etiology. AD.

Comments. Robinson et al [1962] described five patients (three women) over three generations. According to the above authors, Baisch [1931]

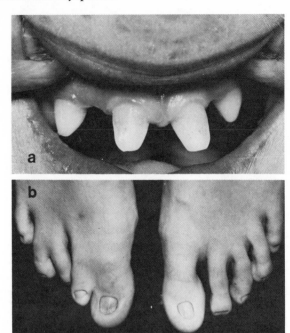

Fig. 11-2. Robinson's syndrome. a. Hypodontia, coniform and widely spaced teeth; note onychodysplasia. b. Syndactyly and onychodysplasia. Courtesy of Dr. Geoffrey C. Robinson, Vancouver, BC, Canada.

reported a girl whose clinical findings were similar to those for the nail, dental, and skeletal abnormalities found in their cases. Pinsky [1975, 1977] and Freire-Maia [1977a] think that they are different entities, suggesting an AR pattern of inheritance for Baisch's syndrome. However, by reading the original paper by Baisch [1931], we verified that there was no consanguinity between the parents of the only affected girl (see below).

Other references. None.

4. BAISCH'S SYNDROME (McK: not listed)

Synonyms. None.
Hair. Normal.
Teeth. Delayed eruption of the central and absence of the lateral incisors.
Nails. Almost total absence of the finger- and toenails.
Sweat. Normal.
Skin. Normal.

Hearing. No data.

Eyes. Normal.

Face. Normal.

Psychomotor and growth development. No data.

Limbs. Polydactyly with syndactyly in the hands (sixth and seventh fingers); hypoplasia of the distal interphalangeal joints of fingers and toes; short and wide hands and feet; adduction of feet.

Other findings. Delayed bone age.

Etiology. Unknown.

Comments. Baisch [1931] described a girl from normal parents. Contrary to the information given by Robinson et al [1962], the patient's parents were not consanguineous. Syndactyly (in the hands in one case, and in the hands and feet in another) and malformed feet were also found in three distant relatives.

Other references. Pinsky [1975, 1977], Freire-Maia [1977a].

5. ODONTOONYCHODYSPLASIA (McK: not listed; see *Comments*)

Synonym. Hereditary hypertrophy of nailbeds associated with erupted teeth at birth.

Hair. Normal.

Teeth. Two or more natal teeth. They decay and disappear at about the sixth to the ninth month.

Nails. Smooth and normal in appearance at the base; at the tip, they are raised by a dark yellowish horny mass of material that projects at an angle from the nailbeds. This angular projection is especially marked in the fingers.

Sweat. Normal.

Skin. Normal.

Hearing. Normal.

Eyes. Normal.

Face. Normal.

Psychomotor and growth development. Normal.

Limbs. Normal.

Other findings. No data.

Etiology. AD.

Comments. Murray [1921] described seven persons (five women) over three generations. They are referred to in McK 16720 as having pachyonychia congenita.

Other references. None.

6. KIRGHIZIAN DERMATOOSTEOLYSIS (McK: not listed)

Synonyms. None.
Hair. Normal.
Teeth. Hypodontia; abnormally shaped.
Nails. Some dystrophic fingernails. See *Comments.*
Sweat. Normal.
Skin. Multiple ulcerations on face, trunk, and limbs, with healing of the more superficial ones and fistulous cicatrization of the deeper ones.
Hearing. Normal.
Eyes. Recurrent keratitis with corneal scarring leading to visual impairment.
Face. Normal.
Psychomotor and growth development. Normal.
Limbs. Acromegaloid enlargement of hands and feet; claw hands; enlarged and deformed joints; short fingers; flexion contractures in some fingers; short left leg leading to scoliosis.
Other findings. Skeletal roentgenograms revealed osteolytic changes, soft tissue swelling and disturbance of growth in length, swelling of metaphyses, and modeling defects of diaphyses of tubular bones.
Etiology. AR?
Comments. Kozlova et al [1983] described five individuals (three women) in a noninbred sibship of 11.

Dystrophic nails may be secondary to osteolysis or nailbed ulceration. This is a mesoectodermal dysplasia—"the Kirghizian form of dermatoosteolysis" (page 210) or "the Kirghizian dermatoosteolysis" (page 205) as the authors call it. Their most severe signs are mesodermic, *not* ectodermic.
Other references. None.

12. Subgroup 2-4

1. NAEGELI-FRANCESCHETTI-JADASSOHN'S DYSPLASIA (McK 16100)

Synonyms. Familial chromatophoric nevus; Naegeli's syndrome; Franceschetti-Jadassohn's syndrome; Naegeli's incontinentia pigmenti; reticular pigmented dermatosis; melanophoric nevus; palmoplantar hyperkeratosis with reticular pigmentation.

Hair. Normal.

Teeth. Carious and yellowish spotted.

Nails. Normal.

Sweat. Hypohidrosis (diminished sweat glands function); discomfort in heat.

Skin. Reticular cutaneous pigmentation (appearing around the age of 2 years and tending to disappear with age), similar to that of the "classical" form of incontinentia pigmenti (Bloch-Sulzberger's syndrome); palmoplantar hyperkeratosis.

Hearing. No data.

Eyes. See *Comments.*

Face. No data.

Psychomotor and growth development. No data.

Limbs. No data.

Other findings. No data.

Etiology. AD.

Comments. Naegeli [1927] described a father and two daughters. Franceschetti and Jadassohn [1954] reexamined Naegeli's patients and found seven other cases in the same family; in one of them, the presence of Fuchs' heterochromia was also noted.

Differences from incontinentia pigmenti include 1) equal frequency in both sexes; 2) palmoplantar hyperkeratosis; 3) absence of blistering and inflammation; 4) absence of ocular disturbances; 5) normal hair; and 6) cutaneous reticular pigmentation appearing at about age 2 (*not* congenital or appearing in the first weeks of life).

Other references. Franceschetti [1953].

2. MARSHALL'S SYNDROME I (McK: not listed; see *Comments*)

Synonyms. Ectodermal dysplasia with ocular and hearing defect; ectodermal dysplasia, deafness and ocular anomalies.

Hair. Normal.

Teeth. Occasional hypodontia, microdontia, protrusion of upper incisors, short roots, abnormalities of eruption, and malposition.

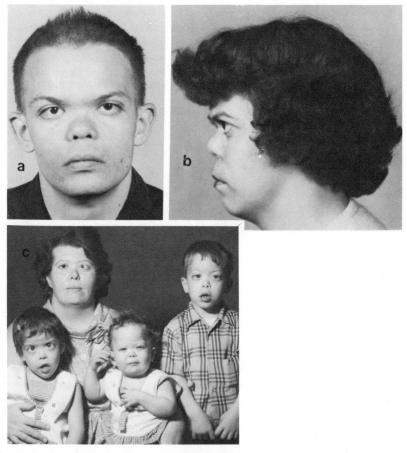

Fig. 12-1.   Marshall's syndrome I. a, b. Hypertelorism, depressed nasal bridge, anteverted nares, malar hypoplasia. c. Affected mother and three children. a, b, courtesy of Dr. Don Marshall, Kalamazoo, MI; c, courtesy of Dr. Jeanne K. Smith, Iowa City, IA.

Nails. Normal.

Sweat. Occasional mild hypohidrosis.

Skin. No data.

Hearing. Progressive congenital sensorineural deficit.

Eyes. Myopia; fluid vitreous; congenital and juvenile cataracts with spontaneous and sudden maturation and absorption; luxation of cataract; esotropia; hypertropia.

Face. Characteristic, with congenital and persistently severe flat nasal bridge, anteversion of nostrils, malar hypoplasia, frontal bossing, and sometimes hypertelorism.

Psychomotor and growth development. Occasional mental retardation; short-to-normal stature.

Limbs. Hypoextensible joints.

Other findings. Cranial and spondyloepiphyseal abnormalities. See *Comments.*

Etiology. AD.

Comments. Marshall [1958] described seven patients (four women) over three generations with some of the above signs, under the heading of "ectodermal dysplasia." He thought that these signs were "suggestive that they [the patients] may represent incomplete examples of hereditary anhidrotic ectodermal dysplasia" (page 143), ie, Christ-Siemens-Touraine's syndrome. Later on, it was verified that tooth defects and hypohidrosis were not constant manifestations of the syndrome [see, for example, Zellweger et al, 1974]; this led Dr. Don Marshall to state: ". . . I really don't believe that my family is an accurate example of ectodermal dysplasia" (letter dated October 15, 1979, to N.F.-M.). Obviously, it is not a "good" example of an ectodermal dysplasia (in the sense that nobody would refer to it as such if a few illustrative examples should be given in a lecture) since dental defects are occasional and hypohidrosis is occasional and doubtful [O'Donnell et al, 1976]. Marshall's syndrome may not even be an ectodermal dysplasia of group A according to the classification of Freire-Maia [1971, 1977a]. However, since it is widely mentioned as an ectodermal dysplasia, we maintain it here only with the intention of discussing the problem. It is possible that it would be better placed in another nosologic group, in spite of the fact that the same condition (such as the Ellis-van Creveld's syndrome) may be classified into two different groups without any clinical or etiological inconvenience.

Contrary to the opinion of Cohen [1974], O'Donnell et al [1976] admit that the Marshall's syndrome I and Stickler's syndrome are not the same entity, since there are differences between them. Some of these differences are the following: 1) Not one of the patients diagnosed as having Stickler's

syndrome has had a thickened calvarium or calcification of falx, tentorium, or meninges; 2) deafness is much less frequent in Stickler's syndrome and is usually conductive; 3) in only one patient with Marshall's syndrome I has there been a cleft palate, whereas about 47% of patients with Stickler's syndrome had cleft palate with or without Pierre Robin's anomaly; 4) Stickler's syndrome has a high incidence of myopia with spontaneous juvenile retinal detachment with the lens in place (the two retinal detachments reported in Marshall's syndrome I were presumably secondary ones, occurring after the lens was removed by cataract surgery); 5) every patient with Marshall's syndrome I has anteversion of nostrils and no one diagnosed as having Stickler's syndrome had it.

Baraitser [1982] reexamined a family originally reported as having a "variant" of Marshall's syndrome I and concluded that its clinical picture encompassed both Marshall's syndrome I and Stickler's syndrome.

Winter et al [1983] described three unrelated patients who had a diagnosis of Weissenbacher-Zweymüller's (W-Z) syndrome at the neonatal period and who later received the diagnosis of Marshall's syndrome I; the authors accept the identity of both. Since both W-Z and Stickler's syndromes have already been found in the same family [Kelly et al, 1982], it is possible that W-Z, Stickler's and Marshall's syndrome I may represent variants of the same syndrome.

It is important to know that there is another Marshall's syndrome described by Dr. R.E. Marshall and co-workers [Marshall et al, 1971] that is rather different from the present one. We call it Marshall's syndrome II.

The paper by Marshall [1958] is referred to in McK 14320 under the heading of "hyaloideoretinal degeneration of Wagner."

Other references. None.

3. AMELOCEREBROHYPOHIDROTIC SYNDROME (McK 22675)

Synonym. Familial epilepsy and yellow teeth.

Hair. Normal.

Teeth. Yellow due to the visibility of dentin as a consequence of enamel hypoplasia in both dentitions.

Nails. Normal.

Sweat. Hypohidrosis (decreased number of sweat glands); sodium and chloride moderately elevated and potassium markedly elevated. See *Comments.*

Skin. Scarce sebaceous glands and nerve fibers.

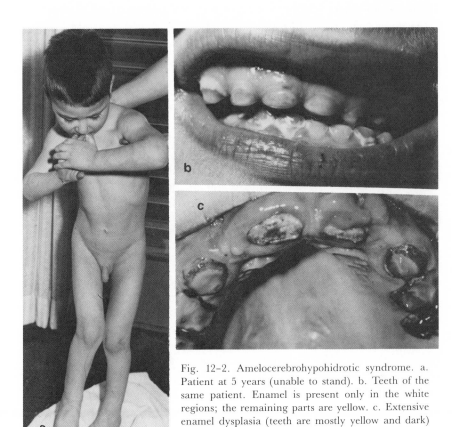

Fig. 12–2. Amelocerebrohypohidrotic syndrome. a. Patient at 5 years (unable to stand). b. Teeth of the same patient. Enamel is present only in the white regions; the remaining parts are yellow. c. Extensive enamel dysplasia (teeth are mostly yellow and dark) in another patient. Courtesy of Dr. Alfried Kolschütter, Göttingen, West Germany.

Hearing. No data.

Eyes. Myopia. See *Comments.*

Face. No data.

Psychomotor and growth development. Progressive CNS degeneration with severe epileptiform seizures appearing between 11 months and 4 years of age; muscle spasticity; abnormal EEG.

Limbs. No data.

Other findings. Brain histology showed diminished number of neurons, small glial cells, ballooning of axons, and lipid-filled pericytes.

Etiology. AR? XR?

Comments. Kohlschütter et al [1974] described five brothers (four deceased) in a sibship of 11 (one woman) from a nonconsanguineous marriage. Sweat abnormalities and myopia were noted in the only living affected man. Kohlschütter [1979, 1981, pers. communications] referred to three other men from Switzerland (one in a sibship of two, and two in a sibship of three; the normal sibs were women). Consanguinity is also denied in these cases. The distance between the birthplaces of the parents of the last two patients is 20 km; the father of the single case was born in the same village as the mother of the other two.

Other references. None.

13. Subgroup 3-4

1. ABSENCE OF DERMAL RIDGE PATTERNS, ONYCHODYSTROPHY, AND PALMOPLANTAR ANHIDROSIS (McK: not listed)

Synonyms. None.

Hair. Normal.

Teeth. Normal.

Nails. Fingernails attached distally to the hyponychium; rough in texture; horizontally and vertically grooved; vertical splits; reduced amount of cuticle. The fingernails bleed easily when fissures or calluses develop around them. In cold weather, the fingernails tend to pull away from the skin.

Sweat. Anhidrosis (lack of sweat pores) on palms and soles.

Skin. At birth, multiple milia (on chin) and several vesicular/bullous lesions (on fingers and soles); they regress later. Leatherlike texture and callous in adults. In cold weather, it becomes very dry.

Hearing. Normal.

Eyes. Normal.

Face. Normal.

Psychomotor and growth development. Normal.

Limbs. Normal.

Other findings. None.

Etiology. AD.

Comments. Reed and Schreiner [1983] described a ten-member (seven women), five-generation pedigree.

Absence of dermal ridge patterns is not a single entity but rather a nosologic group; the other members of this group seem to be clearly different from the present one and are not covered by our definition of ectodermal dysplasia.

Since the authors did not suggest a name for this condition, the above-mentioned one is being suggested by ourselves.

Other references. None.

14. Clinical Overview

Table 14-1 presents the overall distribution of clinical signs among ectodermal dysplasias. For the specific signs, the reader should consult the descriptions given in the previous chapters.

Other things being equal (actual distribution and ascertainment probability), the 11 subgroups should appear with equal frequencies in our sample. However, their numbers show tremendous disparities ($\chi_{10}^2 = 97.08$; $P < 10^{-4}$), from as high as 31 and 25 to as low as 1 and 3. Our impression is that this reflects, at least in part, the real occurrence of ectodermal dysplasias in the general population. Note that the most frequent conditions belong to the subgroups 1-2-3, 1-2-3-4, 1-2, and 1-3; this fact may reflect pathogenetic relationships to be elucidated later on.

Looking at the total of the items (hair, teeth, nails, and sweat), another source of heterogeneity seems evident, since they occur with the respective frequencies of 91, 80, 75, and 42%. Obviously, trichodysplasias call more attention to themselves than, say, dental defects or dyshidrosis; thus, ascertainment through location of patients either by questionnaires sent to local authorities or by visits to communities may reveal an excess of patients with hair problems, whereas ascertainment through medical and odontological clinics will probably discover more patients with more severe symptoms, such as those related to dental and sweat problems. If the two opposed biases balance each other, equality of frequencies would be expected if the subtypes really occur with equal frequencies in the general population. However, the situation is rather different, as shown by the above frequencies. Therefore, it seems reasonable to consider that trichodysplasia is really a common sign (around 90%) while dyshidrosis is the least frequent of them (around 40%).

The frequencies of the other items are as follows: skin (85%), face (72%), other findings (65%), psychomotor and growth development (61%), eyes (60%), limbs (48%), and hearing (24%). It seems reasonable to consider skin alterations as the most frequent (85%) and hearing deficit (either conductive, sensorineural or both) as the less frequent (occurring in about one fourth

183

TABLE 14-1. Subgroups of ectodermal dysplasias of group A with the distribution of their respective clinical signs[a]

Subgroups	Hair	Teeth	Nails	Sweat	Skin	Hearing	Eyes	Face	Psychomotor and growth development	Limbs	Other findings
1-2-3-4	25	25	25	25	25(100)	10(40)	17(68)	18(72)	16(64)	11(44)	21(84)
1-2-3	31	31	31	0	28(90)	4(13)	22(71)	28(90)	17(55)	22(71)	23(74)
1-2-4	6	6	0	6	5(83)	1(17)	4(67)	4(67)	2(33)	2(33)	3(50)
1-3-4	9	0	9	9	9(100)	3(33)	6(67)	7(78)	7(78)	1(11)	3(33)
2-3-4	0	1	1	1	1(100)	0	0	0	0	0	0
1-2	19	19	0	0	12(63)	6(32)	9(47)	15(79)	15(79)	9(47)	16(84)
1-3	12	0	12	0	9(75)	0	5(42)	3(25)	7(58)	3(25)	4(33)
1-4	3	0	0	3	2(67)	0	2(67)	3(100)	1(33)	0	1(33)
2-3	0	7	7	0	4(57)	3(43)	1(14)	3(43)	3(43)	6(86)	2(29)
2-4	0	3	0	3	2(67)	1(33)	3(100)	2(67)	2(67)	1(33)	2(67)
3-4	0	0	1	1	1(100)	0	0	0	0	0	0
Total, 117	105(91)	92(80)	86(75)	48(42)	98(85)	28(24)	69(60)	83(72)	70(61)	55(48)	75(65)

[a]Percentages are shown in parentheses.

of the conditions). The other items seem to occupy an intermediate position. It may be precipitous to advance the idea that frequencies of face involvement, deficit in the psychomotor and growth development, eye involvement, and limb malformations really differ as much as our current data seem to indicate. Further research will be necessary to elucidate this point.

An analysis of the distribution of the items among the different subgroups seems to reveal some differences that are probably real. However, the number of conditions is very low (1, 1, 3, 3, 6, 7, 9) in seven subgroups. Random fluctuations and ascertainment biases could well explain some differences. However, a few cases seem to reflect the real situation in the population at large. For example, among subgroups 1-2-3-4, 1-2, 1-2-3, and 1-3, hearing defects are present as follows: 10/25, 6/19, 4/31, and 0/12 ($\chi_3^2 = 10.23$; P < 0.02). Subgroup 1-3 also seems to present less face involvement (3/12) than the other three (18/25, 15/19, 28/31, respectively) ($\chi_3^2 = 19.37$; P < 10^{-3}). A few other similar situations can also be found. Larger samples will obviously be needed to elucidate this point.

15. Etiological Overview

As shown in the previous chapters, ectodermal dysplasias present a high degree of clinical and causal heterogeneity. The 117 conditions may be causally distributed, as shown in Table 15-1. When stating the cause of each condition, we adopted the following criteria:

1. *Unknown cause.* When the condition is represented by only a single case from one normal nonconsanguineous marriage.

2. *Autosomal recessive?* When the condition is represented by more than one affected in at least one noninbred sibship and/or by a few isolated cases in different small, noninbred sibships.

3. *Autosomal recessive.* When at least one consanguineous marriage is present in the above situations or when the only known sibship is inbred.

4. *Autosomal dominant?* When only transmission from mother to sons and/ or daughters is verified (a situation also compatible with the hypothesis of an X-linked dominant gene) or when in at least one of the parents one or a few mild signs of the condition have been described (in such a case, the gene would have variable expressivity).

5. *Autosomal dominant.* When transmission from both sexes to both sexes (or at least from father to sons and daughters) is present. The presence of a segregation ratio of 1:1:1:1 (with regard to sex and affection) as well as transmission along several generations reinforce this hypothesis.

6. *Autosomal incomplete dominant.* When both parents have mild manifestations of the full picture of the condition seen in their children. A question mark after the notation AID means that the evidence is poor.

7. *X-linked recessive?* When only affected men are known, but this could be due to chance (in such case, the hypothesis of AR inheritance is also possible) or when an affected boy has an apparently affected maternal great-uncle (seen only in photographs). These situations refer to two conditions—amelocerebrohypohidrotic syndrome (AR? XR?) and trichofaciohypohidrotic syndrome (XR?), respectively. However, note that these conditions are listed under the heading "unknown cause," with X-linkage being a mere suggestion.

8. *X-linked recessive*. When the data show transmission of the gene from "normal" mothers (carriers) to affected sons, with affected men on the maternal side only. Normal and affected sons occur at the 1:1 ratio. The best known example is that of Christ-Siemens-Touraine's (CST) syndrome, in which segregation analysis really shows a 1:1 ratio among sons of carriers and about 70% of the carriers may be recognized by the presence of mild and variable signs.

9. *X-linked dominant?* When the pedigrees are compatible with an AD interpretation and the offspring of affected fathers is compatible with XD interpretation. There are neither cases of affected fathers with affected sons and/or normal daughters (this would rule out the XD interpretation) nor of several affected fathers with only affected daughters (this would proportionally reinforce the XD interpretation and, therefore, the question mark would be deleted). Sometimes both AD and XD hypotheses are put together with a question mark.

10. *X-linked dominant*. When affected fathers transmit the condition to all of their daughters, their sons being normal, and affected mothers transmit it to half of their offspring irrespective of sex.

With different degrees of reliability (ie, including and excluding in each causal category the conditions with unknown but suggested cause), the following list summarizes the 117 causal conditions represented, because of heterogeneity, by 108 clinical conditions (Table 15-1) (the corresponding percentages are given within parentheses): Autosomal dominants: 35–44 (29.9–37.6); autosomal recessives: 30–51 (25.6–43.6); X-linked conditions: 5–14 (4.3–12.0); unknown cause: 17–56 (14.5–47.9).

Note that $35 + 30 + 5 + 47 = 117$. The value 56 is replaced by 47(40.2%) because there are nine conditions that are counted twice; although with unknown cause, two different causes are possible for each of them—six AD or XD, one AD or AR, and two AR or XR.

The above results should be understood as follows: Number of conditions with good evidence of AD inheritance, 35; number of conditions with good evidence plus those with less good evidence, 44 (note that this value includes six conditions that may have either an AD or XD inheritance plus one that may be AD or AR). Number of conditions with good evidence of AR inheritance, 30; number of conditions with good evidence plus those with less good evidence, 51 (note that this value includes two conditions that may have either an AR or XR inheritance plus the above-mentioned one with an AD or AR inheritance). Number of conditions with good evidence of X-linked inheritance, 5; number of conditions with good evidence plus those with less good evidence, 14 (note that this value

TABLE 15-1. List of the conditions with indication of cause

Conditions	AD	AR	XL	Heterogeneity AD	Heterogeneity AR	Heterogeneity XL	?	Unknown AD?	Unknown AR?	Unknown XL?
Subgroup 1-2-3-4										
1. Christ-Siemens-Touraine's (CST) syndrome			XR							
2. Autosomal recessive hypohidrotic ectodermal dysplasia		x								
3. Focal dermal hypoplasia (FDH) syndrome								x		XD*
4. Xeroderma-talipes-enamel defect	AID									
5. Rosselli-Gulienetti's syndrome		x								
6. Dyskeratosis congenita				x	x	XR				
7. Pachyonychia congenita	x									
8. Rapp-Hodgkin's syndrome	x							x		XD
9. Ectrodactyly-ectodermal dysplasia-cleft lip/palate (EEC) syndrome	x									
10. Ankyloblepharon-ectodermal defects-cleft lip and palate (AEC) syndrome	x									
11. Zanier-Roubicek's syndrome	x									
12. Trichoonychodental (TOD) dysplasia	x									
13. Jorgenson's syndrome								x		XD
14. Carey's syndrome							x			
15. Camarena syndrome								x		XD
16. Ichthyosiform erythroderma-deafness-keratitis		x								
17. Anonychia with bizarre flexural pigmentation	x									
18. Hypohidrotic ectodermal dysplasia with papillomas and acanthosis nigricans									x	
19. Odontoonychohypohidrotic dysplasia with midline scalp defect	x									
20. Odontoonychodermal dysplasia		x								
21. Trichoodontoonycho-hypohidrotic dysplasia with cataract		x							x	
22. Papillon-Lefèvre's syndrome		x								
23. Hypomelanosis of Ito	x									
Total	9	5	1	1	1	1	1	4	2	1*+3

(continued)

TABLE 15-1. List of the conditions with indication of cause (continued)

Conditions	AD	AR	XL	Heterogeneity AD	Heterogeneity AR	Heterogeneity XL	?	Unknown AD?	Unknown AR?	Unknown XL?
Subgroup 1-2-3										
1. Rothmund-Thomson's syndrome		x								
2. Fischer-Jacobsen-Clouston's syndrome	x									
3. Coffin-Siris' syndrome								x	x	
4. Odontotrichomelic syndrome										
5. Trichodentoosseous (TDO) syndrome I	x									
6. Trichodentoosseus (TDO) syndrome II	x									
7. Trichodentoosseus (TDO) syndrome III	x									
8. Incontinentia pigmenti										XD*
9. Cranioectodermal syndrome								x	x	
10. Fried's tooth and nail syndrome		x								
11. Hypodontia and nail dysgenesis	x									
12. Dentooculocutaneous syndrome							AID			
13. Trichorhinophalangeal (TRP) syndrome I				x	x?					
14. Ellis-van Creveld's syndrome		x								
15. Cystic eyelids—palmoplantar keratosis—hypodontia—hypotrichosis		x								
16. Šalamon's syndrome		x								
17. Trichooculodermovertebral syndrome		x								
18. Oculodentodigital (ODD) syndrome II							x			
19. Arthrogryposis and ectodermal dysplasia							x			
20. Trichoodontoonychodermal syndrome							x			
21. Trichoodontoonychial dysplasia									x	
22. Odontoonychodysplasia with alopecia									x	
23. Schinzel-Giedion's syndrome									x	
24. Growth retardation—alopecia—pseudoanodontia—optic atrophy (GAPO)		x								
25. Ectodermal dysplasia with syndactyly		x								

	C1	C2	C3	C4	C5	C6	C7	C8	C9	C10
26. Osteosclerosis and ectodermal dysplasia		x								
27. Dermoodontodysplasia	x									
28. Trichoodontoonychodysplasia with pili torti			x				x			XD
29. Mesomelic dwarfism—skeletal abnormalities—ectodermal dysplasia			x							
30. Ectodermal dysplasia syndrome with tetramelic deficiencies										
Total	6	9	0	1	1?	0	5	4	6	1*+1
Subgroup 1-2-4										
1. Regional ectodermal dysplasia with total bilateral cleft							x			
2. Melanoleucoderma		x								
3. Böök dysplasia	x									
4. Congenital insensitivity to pain with anhidrosis		x								
5. Lenz-Passarge's dysplasia			XD							
6. Ectodermal dysplasia with palatal paralysis										
Total	1	2	1	0	0	0	2	0	0	0
Subgroup 1-3-4										
1. Fischer's syndrome	x									
2. Trichodysplasia—onychogryposis—hypohidrosis—cataract							x			
3. Alopecia—onychodysplasia—hypohidrosis—deafness							x			
4. Hayden's syndrome							x			
5. Alopecia—onychodysplasia—hypohidrosis							x			
6. Hypohidrotic ectodermal dysplasia with hypothyroidism										
7. Ectodermal dysplasia with severe mental retardation							x		x	
8. Alopecia universalis—onychodystrophy—total vitiligo										
9. Dermotrichic syndrome			XR						x	
Total	1	0	1	0	0	0	5	0	2	0
Subgroup 2-3-4										
1. Ameloonychohypohidrotic dysplasia	x									
Total	1	0	0	0	0	0	0	0	0	0
Subgroup 1-2										
1. Orofaciodigital (OFD) syndrome I			XD	x						
2. Oculodentodigital (ODD) syndrome I				x						

(continued)

TABLE 15-1. List of the conditions with indication of cause (continued)

Conditions				Heterogeneity			Unknown			
	AD	AR	XL	AD	AR	XL	?	AD?	AR?	XL?
3. Hallermann-Streiff's syndrome				x?	x					
4. Gorlin-Chaudhry-Moss' syndrome									x	
5. Mikaelian's syndrome		x								
6. Gingival fibromatosis—sparse hair—malposition of teeth		x								
7. Hypertrichosis and dental defects	x									
8. Gingival fibromatosis and hypertrichosis				x	x?					
9. Pili torti and enamel hypoplasia	x									
10. Walbaum-Dehaene-Schlemmer's syndrome		x								
11. Oculoosteocutaneous syndrome									x	
12. Agammaglobulinemia—thymic dysplasia—ectodermal dysplasia							x			
13. Johanson-Blizzard's syndrome		x								
14. Trichodental dysplasia	x									
15. Acrorenal-ectodermal dysplasia—lipoatrophic diabetes (AREDYLD) syndrome		x								
16. Alopecia—anosmia—deafness—hypogonadism	x		x							
Total	4	5	1	2+1?	2+1?	0	1	0	2	0
Subgroup 1-3										
1. Hairy elbows dysplasia		x								
2. Palmoplantar hyperkeratosis and alopecia				x						
3. Curly hair—ankyloblepharon—nail dysplasia (CHANDS)		x			x?					
4. Onychotrichodysplasia with neutropenia		x								
5. Pili torti and onychodysplasia	x									
6. Agammaglobulinemia—dwarfism—ectodermal dysplasia									x	
7. Trichoonychodysplasia with xeroderma		x								
8. Skeletal anomalies—ectodermal dysplasia—growth and mental retardation							x			
9. Sabinas brittle hair and mental deficiency syndrome		x								
10. Syndrome of accelerated skeletal maturation, failure to thrive and peculiar face									x	
11. Congenital lymphedema, hypoparathyroidism, nephropathy, prolapsing mitral valve and brachytelephalangy									x	XR
Total	1	5	0	1	1?	0	1	0	3	1

									XR
Subgroup 1-4										
1. Ectodermal dysplasia of the head	0	0	0	x	x?	0	0	0	x	1
2. Trichofaciohypohidrotic syndrome										
Total	0	0	0	1	1?	0	0	0	1	0
Subgroup 2-3										
1. Triphalangeal thumbs—onychodystrophy—deafness			x	x?	x?		x	x		
2. Ectodermal defect with skeletal abnormalities										
3. Robinson's syndrome	x									
4. Baisch's syndrome							x			
5. Odontoonychodysplasia	x									
6. Kirghizian dermatoosteolysis									x	
Total	2	0	1	1	1	0	2	1	1	0
Subgroup 2-4										
1. Naegeli-Franceschetti-Jadassohn's dysplasia	x									
2. Marshall's syndrome I	x								x	x
3. Amelocerebrohypohidrotic syndrome									x	1
Total	2	0	0	0	0	0	0	0	1	1
Subgroup 3-4										
1. Absence of dermal ridge patterns, onychodystrophy and palmoplantar anhidrosis	x									
Total	1	0	0	0	0	0	0	0	0	0
Subgroups (No. of conditions)										
1-2-3-4 (23)	9	5	1	1	1	1	1	4	2	1*+3
1-2-3 (30)	6	9	0	1	1?	0	5	4	6	1*+1
1-2-4 (6)	1	2	1	0	0	0	2	0	0	0
1-3-4 (9)	1	0	1	0	0	0	5	0	2	0
2-3-4 (1)	1	0	0	0	0	0	0	0	0	0
1-2 (16)	4	5	1	2+1?	2+1?	0	1	0	2	0
1-3 (11)	1	5	0	0	1?	0	1	0	3	1
1-4 (2)	0	0	0	1	1?	0	0	0	0	1
2-3 (6)	2	0	0	1	1	0	2	0	1	0
2-4 (3)	2	0	0	0	0	0	0	0	1	1
3-4 (1)	1	0	0	0	0	0	0	0	0	0
Grand total (108)	28	26	4	7+1?	4+4?	1	17	8	17	2*+7

[a]The asterisk indicates the most probable of two possible causes. Sometimes the number of conditions of the clinical subgroups do not correspond to the addition of the values under each causal subgroup because of heterogeneity and the existence of more than one causal possibility (see text). AD, autosomal dominant; AR, autosomal recessive; XL, X-linked.

includes eight conditions that may have either an AD or XD inheritance (six) and an AR or XR inheritance (two). For the conditions with unknown cause, the first value (17) represents the conditions with no suggestion at all of a possible cause and the second (56) the number of conditions with a totally unknown cause (17) plus those with mere suggestions of a cause (39).

These data show that when the genetic hypotheses are fully supported by the data, AD and AR conditions are similarly represented (around 28%), whereas X-linked conditions appear at the lowest frequency (4.3%). With such a degree of reliability, 40.2% of the conditions are of presently unknown cause. When, however, reliability decreases and mere suggestions are accepted as supporting the hypotheses, ARs appear in first place, at 43.6%; ADs are second (37.6%); X-linked conditions increase from 4.3 to 12.0%; and "unknown causes" decrease from 40.2% to 14.5%.

Autosomal dominant conditions are much more easily ascertained and investigated than AR conditions. However, AR and AD conditions are equally frequent among those having a reliable cause, and AR conditions outnumber AD ones in the total. This suggests that an AR mechanism of inheritance actually may be more common than an AD mechanism among patients in general. Since an X-linked mechanism is easier to ascertain than an AR mechanism, it is possible that its frequency in our sample of ectodermal dysplasias corresponds roughly to reality if it is not an overestimate.

No attempt will be made to analyze statistically the distribution of causes among the subgroups, not only because some of them present low numbers of conditions (this fact did not prevent us from performing a similar analysis in the previous chapter), but especially because the cases of unknown causes (with or without mere suggestions of cause) and of possible but doubtful heterogeneity introduce an uncertainty into the data. We would simply like to call attention to the fact that AD inheritance predominates in subgroup 1-2-3-4, whereas the AR inheritance seems to be more common in subgroups 1-2-3 and 1-3. These differences may represent mere sample bias, but it is possible that with an increase in the number of conditions, such apparent differences may prove to be real.

Not all the etiological information without a question mark has the same degree of reliability. For example, cause is firmly established in the Christ-Siemens-Touraine's syndrome (XR), but only suggested in trichooculo-dermovertebral syndrome (AR) and gingival fibromatosis, sparse hair, and malposition of teeth (AR).

As shown above, among ectodermal dysplasias, there are a number of conditions whose cause is not well established. Since many conditions are still inadequately known at the clinical, physiological, metabolic, or radiographic level, they generally need reinvestigation at all levels. It is noteworthy that "new" ectodermal dysplasias are being described each year, which means that the field is also rewarding for discovering hitherto-undescribed conditions.

16. Genetic Counseling

INTRODUCTION

Counseling should not be confused with information transfer, which transmits data, risks, hypotheses, classifications, etc, in an impersonal way, either privately or publicly. If someone asks about the risk of giving birth to a second affected child in a given sibship, he is asking for mere information. He may be personally interested in the answer either because he is the father, the uncle, the cousin, or the neighbor of that sibship or simply because he heard about the condition and wants to know more about it. If he becomes satisfied with the short and cold answer (say, a 25% risk), or asks for a more detailed explanation, he is still asking for mere information. Indeed, the informant may never even know why the question was posed to him.

Counseling is much more than this. A textbook on human genetics may give a detailed analysis of the main problems of the field without saying a word on *how* to provide counseling, since this activity requires much more than knowledge and capacity to transmit it. During a counseling session, profound emotional problems may emerge whether the counselor is well prepared for the task or not. That is why counseling is much more time-consuming than mere information transfer. Having worked for more than 25 years in this field, the senior author has observed that, at the beginning of his activities in this area, he used to provide much more information than counseling. Only after some years did he discover the profound differences between the two activities, and thus was able to begin doing his work in a correct way. For instance, after discovering those profound differences, he realized that, in some situations, good counseling requires sessions of no less than 1 hour.

A busy counseling service, providing a number of consultations per day or concentrating all of them on one given day per week, obviously cannot provide counseling in the way we believe it should be done. Generally, in such cases, counseling becomes information transfer: it is short, rapid, and distant; not a long talk during which emotional problems (related to fear, religion, moral doubts, family problems, guilt feelings, adoption of children, etc) may arise and then have to be dealt with. In this context, questions that are asked frequently and are not specifically related to the biological aspect of the situation include: How about marrying my cousin? If I do so, can we have children? Should we try again if the recurrence risk is 25%? Which contraceptive method is best? Do you think that the use of contraceptives is a sin? Do you think I ought to have my tubes tied? What is your opinion on our desire to accept the risk and have another baby? Is abortion a crime? Is it a sin? What do you suggest we do? Is it possible to untie my tubes? Do you know some physicians who do this? Suppose the diagnosis given in the case of our child is not correct; who would know better? Do you think it would be advisable to see another specialist? Which one do you suggest? I think that my child's condition came from my husband's family, not from mine, which is free of such diseases; what do you think? If the average risk is low, in what sense can I trust this average? Would you? What do people generally do in our situation? Do you know any great Swedish specialist in this field? Could you ask the opinion of an American doctor? If you were in love with a cousin, would you marry her? A doctor told me that Huntington's chorea is due to an autosomal recessive gene; my father, his father, and some of his sibs were affected and I thought I was practically safe; now I read in a newspaper that the disease is due to an autosomal dominant gene; so I have a very high probability of becoming affected! You said that this risk is an empirical one; why? Does it apply to my situation or is it just a guess?

RISKS

The probability of any pregnancy resulting in a defective child depends on a number of environmental and genetic variables: The type of marriage (consanguineous or not); a family history of malformations or anomalies; fetal, uterine, and placental factors; the circumstances of labor; presence or absence of polyhydramnios or olygohydramnios; the occurrence of certain infectious diseases, the use of certain drugs and exposure to ionizing radiation during the first trimester of pregnancy; parental (especially maternal) age; parity; twinning; maternal diabetes; etc.

The table of recurrence risks encompasses a range that goes from close to 0 (for example, the majority of deficiency anomalies of the limbs) to close to 1 (when both parents are homozygous for the same allele). However, these extreme situations represent a minority out of the total number of consultations at a genetic counseling service. Many of the risks are 50% (for fully penetrant AD traits), 25% (for AR genes), and lower (15, 10, 4, 2%...) for conditions due to multifactorial determination, to different genetic mechanisms leading to such averages, and finally to unknown causes [for examples, see Freire-Maia and Freire-Maia, 1964, 1967a, b; Freire-Maia and Azevedo, 1968, 1977; Arce et al, 1968; Freire-Maia, 1969, 1975a, 1980; Arce-Gomez et al, 1970; Freire-Maia et al, 1973].

Three categories of risks exist: 1) those based on the cause (in our case, they are called genetic risks); 2) those based on averages of recurrences observed in different sibships with similar conditions (empirical risks); and 3) those due to exogenous factors such as toxoplasmosis, rubella, diabetes, thalidomide, but for which there is no constant relationship between the presence of the factor in the mother and the presence of the syndrome in the child (semi-empirical risks). (For an analysis of the problem and for the methods of estimating empirical risks, see Freire-Maia, [1970b]). The risks for the offspring of consanguineous matings may also be called semi-empirical since we know that consanguinity is the cause of the risk but there is not a constant relationship between the presence of this factor and the occurrence of an increased mortality or morbidity in the offspring [cf. Freire-Maia, 1984].

Since many and different biases affect ascertainment, and empirical risks are only averages of different risks operating in different families, such risk figures should be provided to consultants with great care. Persons seeking consultation should be cautioned about the tentative nature of these risk estimates and told clearly that they describe the average situation in a sample of families, and not the case in their own family. Therefore, the formulation of the advice must be different according to whether we are using an empirical risk figure or a probability based on cause. We may say to the normal consanguineous parents of a child with an AR anomaly that the recurrence probability per pregnancy is 25%; however, it is not equally advisable to tell the normal parents of a child with, say, cleft lip and palate, that the probability of the birth of another child equally affected or affected with only cleft lip is around 4% per pregnancy. In this last case, it is always necessary to state that the figure given is an empirical one, that it is simply a weighted mean in a sample of different sibships with children presenting the same clinical entity, and that in the present case the risk may be higher or lower. In the first example, the recurrence risk really *is* 1/4; in the second it is *not* 4%. A final example: If I tell a normal, nonconsanguineous couple, without any history of malformations in the family, that the recurrence risk of a deficiency anomaly of the limbs *is* negligible (because the data I have in hand give me an average negligible risk), the couple may be upset at the birth of another similarly affected child. Unbeknown to them, the couple could be distantly related and the malformation may have been due to an AR gene; I thus did not know that in that specific situation the risk was 25%.

ECTODERMAL DYSPLASIAS

A nosologic group is an array of conditions that have been pooled on the basis of two general criteria: similarity and convenience. Similarity may be purely clinical (as is the case for ectodermal dysplasias; Freire-Maia, [1971, 1977a]) or pathogenetic (for example, the mucopolysacchari-doses). When clinical similarity is so profound that pathogenetic similarity is likely, a better concept is achieved—that of the "disease communities" or "families of conditions" [Pinsky, 1974, 1975, 1977; Freire-Maia, 1977b]. Obviously, etiological heterogeneity is expected to occur in any of these groups, since similarity of phenotypes does not indicate similarity of geno-types. In other words, similar anomalies may have different causes and thus potentially different recurrence risks.

Ectodermal dysplasias form a very heterogeneous group of conditions; therefore, no general rule can be given regarding recurrence risks. More than that, this group includes a large number of conditions that are poorly known both clinically and causally. Diagnosis may be difficult and in about 40% of all situations our knowledge regarding cause is nil or based on mere conjecture. This means that, in about half the conditions, our coun-seling may be unreliable. And it is sad to say that in the other half we still face some uncertainties.

Autosomal Dominant Conditions

In such situations, if penetrance is complete, affected individuals have an affected parent, women and men are equally affected, and in the total offspring of a sample of couples with one affected partner (including the matings with only normal offspring), there are about 50% normal and 50% affected individuals.

If one of the parents is affected (with or without affected children), then the risk is 50% for each pregnancy. Normal persons in the family (even sibs of affected individuals) married to normal persons have the same probability (equal to twice the extremely low mutation rate) of having affected children as any other normal persons in the population. If penetrance is incomplete, normal persons in the family may have the gene without manifesting its effect, with a consequent probability (that should be calculated for each condition) of having affected children much closer to 50% than to twice the mutation rate. Among ectodermal dysplasias, segregation analysis did not reveal any case of reduced penetrance. However, this does not mean that penetrance is always complete among them.

One of the genetically well-studied AD dysplasias that showed complete penetrance is Fischer-Jacobsen-Clouston's syndrome (1-2-3 subgroup). Thus far, only one pedigree is known of Zanier-Roubicek's syndrome (1-2-3-4 subgroup), but 21 affected individuals (11 men) are present in it and there is direct transmission from one generation to the next (including from father to son), without skipping. The same happens to dermoodontodysplasia (with 11 affected; 1-2-3 subgroup) and to absence of dermal ridge patterns, onychodystrophy, and palmoplantar anhidrosis (with ten affected; 3-4 subgroup), among others. These situations illustrate the fact that it is easier to find and to study anomalies due to AD inheritance. A single pedigree can firmly support this hypothesis through a simple but reliable segregation analysis, while it would be generally insufficient to support AR, XR, or XD inheritance. The frequency of conditions due to AD genes in some samples may reflect this bias.

When the effect manifested by the mutant gene in heterozygotes is constant, easily verifiable, and represents a part of the full picture described in the homozygotes, it is convenient to call the gene incompletely dominant. This seems to be the case for the XTE syndrome (1-2-3-4 subgroup). Heterozygotes have an enamel defect, whereas homozygotes have the complete syndrome, with anomalies of hair, teeth, nail, sweating, eyes, etc.

Coffin-Siris' syndrome (1-2-3 subgroup) may be due to a dominant gene with variable expressivity, since three fathers of sporadic cases had mild signs. Unfortunately, the pedigrees are small and segregation analysis does not support the interpretation firmly. Reduced penetrance may also obscure interpretation if there are insufficient data.

Autosomal Recessive Conditions

In such instances, the parents are normal, have a higher frequency of consanguinity than that prevailing in the population in which they were born, men and women are equally affected, the offspring of an affected married to a nonconsanguineous normal party generally are normal, and, in the total of sibships with at least one affected, the ratio of normal to affected individuals is about 3:1 after appropriate correction for ascertainment bias. The recurrence risk is, therefore, 1/4 (or 25%).

This risk applies to conditions firmly established to be due to an AR gene. However, we listed some conditions as "AR?." This means that the cause is unknown but the genetic interpretation is *possible* (see Chapter 15). In such cases, the couple seeking counseling should be informed that the cause is unknown but that an AR mechanism is *possible* and that, therefore, the recurrence risk may be as high as 25% per pregnancy.

Normal sibs of affected children with AR dysplasias have a high probability (2/3 or 67%) of being heterozygotes (Aa). There is also a high probability that normal cousins are equally heterozygotes. For first cousins, this probability is 1/4; for first cousins once removed it is 1/8; for second cousins it is 1/16; etc. If a normal sib marries a normal cousin, the respective probabilities of having an affected child are 1/24, 1/48, 1/96, etc, per pregnancy. These probabilities decrease as the number of normal children increases without the appearance of an affected individual. However, if an affected child is born, the recurrence risk will be 1/4 for any of the situations listed above. The birth of an affected child (aa) is proof that both parents are heterozygotes (Aa) and therefore we no longer need to multiply different probabilities to estimate the resultant probability of both being heterozygotes.

It is possible that the person being counseled in the above situation may ask, "Before I married, you told me that my risk of having an affected child was (say) 1/24, that is, only about 4%. Now you have changed the probability and tell me that for my next child it is 1/4, that is, 25%. So, probabilities do change; you increased it now!"

The answer is simple. It was not the probability, but our knowledge of it that changed. Before the marriage, it was 1/4 because both were heterozygotes, *but we did not know it.* We only knew it after the birth of an affected child. However, the probability of 1/4 is not the final one. If ultrasonography or biochemical analysis can reveal the defect prenatally, the probability will change either to 1 (affected) or to zero (normal). The probability of 1/4 is also an average of 3/4 of zeros and 1/4 of ones.

If the sib of an affected person marries a nonrelative, the probability of having an affected child is negligibly small, since, despite the 2/3 probability of the individual being a heterozygote, the gene is rare and the probability that he or she would marry a heterozygote at random from the population is very small.

As regards uncles and aunts, their probability of being heterozygotes is 1/2. If they marry their cousins, obviously there will be a high probability (1/16) of both parties being heterozygotes; if they marry nonrelatives, the risks turn out to be negligible.

TABLE 16-1. Risks of the birth of a child with an autosomal recessive condition[a]

	Possible marriage partners			
	1st cousin	1½ cousin	2nd cousin	Nonrelated
The affected	1/8	1/16	1/32	Negligible
Sib	1/24	1/48	1/96	Negligible
Parent	1/4	1/4	1/4	1/4
Uncle/aunt	1/64	1/128	1/256	Negligible
First cousin	1/64	1/128	1/256	Negligible

[a]The risks to a first cousin of a homozygote who has married a cousin of the three highest degrees (1, 1½, and 2) apply only if these cousins are also related to the homozygote with the same degree of consanguinity. They also apply to two cousins of the affected even if they are not cousins themselves. However, the first risk for a first cousin does not include a marriage with a homozygote's sister or brother, in which case the risk is 1/24.

Table 16-1 summarizes some of the risks of the birth of an affected child. This table should be read as follows (on the basis of some examples): If an affected individual marries a second cousin, the risk of having an affected baby is 1/32; if his sister marries a first cousin, the risk is 1/24; for his parents, the risk is always 1/4, because, whether consanguineous or not they are certainly heterozygotes; if one of his uncles marries a first cousin (of the uncle), the risk is 1/64; and, finally, if a first cousin of the affected marries another first cousin of the affected (whether they themselves are cousins or not), the risk is 1/64.

Ellis-van Creveld's syndrome (1-2-3 subgroup) is one of the few for which segregation analysis is available. From the data reported by Pilotto [1978] and compiled from the literature, we found 391 normal and 203 affected (34.2%) individuals among the total of the 133 sibships with at least an affected individual. This high frequency reflects ascertainment bias. Assuming truncate selection (this is not exactly what happened, but it does not seem to be too far from reality, due to the high contribution of the Amish data of McKusick et al [1964]), Pilotto [1978] verified that the ratio obtained after correction did not depart significantly from 3:1—thus confirming AR inheritance. Some of the heterozygotes show a few signs (polydactyly, congenital heart defect, different bone dysplasias, synostosis hamate-capitate, syndactyly, etc; McKusick et al, [1964]; Pilotto [1978]); thus it is incompletely recessive.

The rate of incomplete recessiveness has been estimated for some species (including man). Human data show estimates generally ranging from 1 to about 10%, but the high values are almost certainly due to sample vagaries; the estimates tend to concentrate around 2-3% [Morton et al, 1956; Kimura, 1959, Salzano et al, 1962; Marçallo et al, 1964; Freire-Maia et al, 1964; Conterio and Barrai, 1966; Freire-Maia, 1975b]. For detrimental genes, this figure means that the adaptive value (Darwinian fitness) of the heterozygotes is decreased on the average by about 2–3% in comparison with normal homozygotes.

Strictly speaking, "recessive" genes—ie, genes with a rate of dominance equal to zero—probably do not exist. After the investigation of a larger number of heterozygotes and/or the application of special techniques of analysis, many genes classified as such have been shown to have a small amount of dominance; thus, by definition, these genes could be called incomplete recessives. However, when their effect is very small in relation to the full picture in the homozygotes, they continue to be called recessive, with the additional information that at least a certain fraction of the carriers can be detected. (For the manifestation in a heterozygote [absence of a thumb] of a *recessive* gene that, in the homozygous state, is responsible for a rather severe anomaly of both upper limbs, see Freire-Maia [1975c]; cf Freire-Maia et al [1959]).

The situation in which normal parents (even when nonconsanguineous) have several normal and affected sons and daughters suggests AR inheritance, as is the case, for example, for the odontotrichomelic syndrome [Freire-Maia, 1970a]. Some authors identified this syndrome with ectrodactyly–ectodermal dysplasia–cleft lip/palate and suggest that the presence of four affected sibs of both sexes in a sibship of eight from normal nonconsanguineous parents could be due to a gonial AD mutation. However, the first hypothesis seems to be the most plausible one.

X-Linked Conditions

X-linked genes are sometimes called dominant, incompletely dominant, or recessive. The first term means either that heterozygous women present the trait in the same way as hemizygous men or that heterozygous women present the trait and hemizygotes are lethal; the second means that hemizygotes (men) present the full anomaly and heterozygotes (women) present mild signs of it; the third means that heterozygotes (women) are clinically normal and only men present the anomaly. However, because of the inactivation of one of the two X chromosomes at the beginning of embryogenesis (in women), no sex has both alleles functioning and these terms cannot be applied properly to X-linked genes. Since the process of inactivation of one of the X chromosomes occurs at random, practically no heterozygous woman has exactly 50% of her cells with each of the alleles functioning. Those with more than 50% of the "abnormal" allele will have a higher probability of showing the trait. If all heterozygous women have the trait, this means that the biochemical activity of the defective gene is sufficiently powerful to produce it even when the gene is in an inactivated state in more than 50% of the cells. If only a certain fraction of heterozygous women have the trait, this means that a "large" number of "defective" cells is necessary for the entire organism to reveal the trait.

It is necessary to remember that heterozygous women may manifest only some of the signs of an X-linked recessive condition. This means that, after careful examination, after the discovery of more sensitive tests, after the application of discriminating statistical analyses, etc, conditions previously categorized as "recessive" should be more properly called "incomplete recessive." In spite of this, incompletely recessive genes (autosomal or X-linked) continue being called "recessive" for the sake of convenience. This should not obscure the fact that recessive alleles probably always have a small amount of dominance.

Only one case of well-established X-linked recessive inheritance can be found among the ectodermal dysplasias: Christ-Siemens-Touraine's syndrome. As mentioned earlier, Pinheiro and Freire-Maia [1979c] and Pinheiro et al [1981a] estimated the probability of carrier detection on the basis of a superficial physical examination of obligate carriers to be about 70%. Nakata et al [1980] found that, by measuring the teeth of heterozygous women, they could "usually" separate them from the other groups (affected males, possible heterozygotes, normal males and mothers of simplex cases). However, this cumbersome procedure does not increase the probability of carrier detection [Freire-Maia and Pinheiro, 1982a].

Since carrier women are the source of affected individuals, our suggestion is that all girls belonging to families with Christ-Siemens-Touraine's syndrome should be examined for the mild signs that can be seen among heterozygotes: hypotrichosis, hypodontia, microdontia, mosaic patchy distribution of sweat pores and body hair, hypohidrosis, etc.

Noncarrier women have a negligible probability (equal to the mutation rate of the gene) of having affected sons. Carriers have a 25% probability of having affected sons and 25% of having carrier daughters (obviously, the remaining 50% contain noncarrier [normal] daughters and normal sons).

Assuming that the the above-mentioned examination of putative Christ-Siemens-Touraine's syndrome carriers is performed in sibships where the frequency of carriers is around 50% and some of the women are identified as carriers and others as possible noncarriers, a possibility remains that at least some of the latter really are carriers (30% of them may be missed on examination). The probability among putative carriers of a given woman's being a nonrecognized carrier is 30% × 50% = 15%. Since the probability of a carrier's having an affected son is 25%, the risk for a specific woman in that condition is about 4%. However, if her first child is a boy with Christ-Siemens-Touraine's syndrome or a girl showing a few mild signs of this condition, then the risk that we would inform her would be 25%. It is not the risk but out knowledge of it that changed. This probability holds true for each event (*per pregnancy*), *not* for multiple events along entire sibships. Therefore, the probability of giving birth to, say, three affected in a sibship of eight has to be calculated as would any other situation.

Incontinentia pigmenti is a condition whose gene has been hypothetically assigned to the X chromosome. Since heterozygous women have the condition and hemizygous males generally die in abortions, the probability of an affected women's having affected daughters or affected (lethal) sons is, for each situation, equal to 25% per pregnancy. The remaining 50% encompass both normal sons and daughters. However, in effect, these women's offspring are composed of 1/3 normal boys, 1/3 affected daughters, 1/3 normal daughters. The alternative hypothesis for the cause of incontinentia pigmenti is that of an AD male-lethal gene. This hypothesis fits the available data and the risks remain the same. The reason the X-linked hypothesis is more attractive is that the condition and survival of heterozygous women could be due to their having only 50% of their cells with the defective gene in active state while male-lethality would be a result of the fact that 100% of their cells contain the gene. There are no reasons to think that an autosomal gene could imitate an X-linked one in such a way. The same reasoning applies to focal dermal hypoplasia (Goltz-Gorlin's syndrome).

The distinction between XD and AD modes of inheritance is that, in the first place, affected men do not have affected sons, whereas in the second they produce affected sons and daughters in equal numbers. If only a small pedigree is available, it may be impossible to discriminate between the two hypotheses [Lenz, 1963; page 79]. However, Settineri [1974] and Settineri et al [1976] analyzed a large Brazilian pedigree with a number of persons with Lenz-Passarge's dysplasia, with affected women segregating the gene equally to sons and daughters but affected men segregating the gene only to daughters. The penetrance of the gene was estimated to be about 87%.

Unknown Cause

Ectodermal dysplasias due to unknown or doubtful causes are to be dealt with carefully since "unknown cause" may simply result from the fact that only one or a few cases have been investigated. If the affected individuals are simplex cases and did not reproduce, each affected may be the result of a de novo dominant mutation. In such a case, the recurrence risk for sibs is negligible. However, if at least one sibship contains two or more affected individuals and some simplex cases are also known, the trait may be an AR one. If the few affected individuals described in the literature are all men, X-linked inheritance is possible. Therefore, "unknown cause" may mean several different things, and the respective risks may range from negligibly small to as high as 25%. The causal suggestions that come from very limited data are to be accepted with the utmost caution.

In the cases of unknown cause with *possible* (even remotely possible) AD, AR or X-linked inheritance, no counseling should be given that only states that the cause is unknown and that the recurrence risk is probably negligible. Attention should be called to the possibility that the risk may be higher than "negligible." Those being counseled should be informed of whether that possibility is "probable" or remote. Obviously, due to the unsatisfactory situation found in about 40% of the conditions in this book, the counselor will generally be floating on an ocean of uncertainty. But that is still better than sinking in a pond of total ignorance.

References

Aceves-Ortega R, Madrigal LR (1977): Displasia ectodermica hidrotica. Dermatologica 21:12-21.

Ackerman JL, Ackerman AL, Ackerman AB (1973a): A new dental, ocular and cutaneous syndrome. Int J Dermatol 12:285-289.

Ackerman JL, Ackerman AL, Ackerman AB (1973b): Taurodont, pyramidal and fused molar roots associated with other anomalies in a kindred. Am J Phys Anthropol 38:681-694.

Agostinelli O (1970): La condrodisplasia ectodermica di Ellis-van Creveld. Gazz Int Med Chir 75:131-153.

Airenne P (1981): "X-Linked Hypohidrotic Ectodermal Dysplasia in Finland. A Clinical, Radiographic and Genetic Study." Academic dissertation, University of Helsinki, Helsinki, Finland.

Aita JA (1969): "Congenital Facial Anomalies With Neurologic Defects. A Clinical Atlas." Springfield: C C Thomas.

Alcântara-Silka M (1977): "Estudos sobre Sudorese em uma Amostra de Caucasóides de Curitiba Classificada por Faixa Etária e Sexo." MSc thesis, Federal University of Paraná, Curitiba, Brazil.

Alexander WN, Allen HJ (1965): Hereditary ectodermal dysplasia in three brothers. Oral Surg 20:802-809.

Alves AFP, Santos PAB dos, Castelo-Branco-Neto E, Freire-Maia N (1980): Um novo síndrome autossômico recessivo de displasia ectodérmica e malformacão. Ciênc Cult 32 (suppl): 679.

Alves AFP, Santos PAB dos, Castelo-Branco-Neto E, Freire-Maia N (1981): An autosomal recessive ectodermal dysplasia syndrome of hypotrichosis, onychodysplasia, hyperkeratosis, dwarfism, kyphoscoliosis, cataract and other manifestations. Am J Med Genet 10:213-218.

Andersen H, Pindborg JJ (1947): A case of total "pseudo-anodontia" in combination with cranial deformity, dwarfism, and ectodermal dysplasia. (In Danish.) Odontol T 55:484-493.

Anderson J, Cunliffe WJ, Roberts DF, Close H (1969): Hereditary gingival fibromatosis. Br Med J 3:218-219.

Andreev VC, Stransky L (1979): Hairy elbows. Arch Dermatol 115:761.

Antley RM, Shields ED, Rosenberg GL, Bixler D (1976): Hypohidrosis with sparse hair, short stature and normal teeth and nails. Birth Defects XII(6):136.

Appel B, Messina S (1942): Pili torti hereditaria. New Engl J Med 226:912-915.

Arbisser AI, Scott CI, Howell RR, Ong PS, Cox HL, Jr (1976): A syndrome manifested by brittle hair with morphologic and biochemical abnormalities, developmental delay and normal stature. Birth Defects XII(5):219–228.

Arce B, Azevedo JBC, Freire-Maia N, Chautard EA (1968): Frequëncias e riscos de recorrência de fissuras labiopalatinas. Rev Paul Med 72:239–246.

Arce-Gomez B, Azevedo JBC, Chautard EA, Freire-Maia N (1970): A genetic study on palatolabial defects. Hum Hered 20:580–589.

Arias S (1971): Genetic heterogeneity in the Waardenburg syndrome. Birth Defects VII(4):87–101.

Arias S (1980): Waardenburg syndrome—two distinct types. Am J Med Genet 6:99–100.

Baisch A (1931): Anonychia congenita, Kombiniert mit Polydaktylie und verzögertem abnormen Zahndurchbruch. Dtsch Z Chir 232:450–457.

Baraitser M (1982): Marshall/Stickler syndrome. J Med Genet 19:139–140.

Baraitser M, Hodgson SV (1982): The Johanson-Blizzard syndrome. J Med Genet 19:302–310.

Bart BJ, Gorlin RJ, Anderson VE, Lynch FW (1966): Congenital localized absence of skin and associated abnormalities resembling epidermolysis bullosa. Arch Dermatol 93:296–304.

Bartlett RC, Eversole LR, Adkins RS (1972): Autosomal recessive hypohidrotic ectodermal dysplasia: Dental manifestations. Oral Surg 33:736–742.

Bartsocas CS, Tsiantos AK (1970): Mental retardation with absent fifth fingernail and terminal phalanx. Am J Dis Child 120:493–494.

Bartsocas CS, Weber AL, Crawford JD (1970): Acrocephalosyndactyly type 3: Chotzen's syndrome. J Pediatr 77:267–272.

Basan M (1965): Ektodermale Dysplasie. Fehlendes Papillarmuster, Nagelveränderungen Vierfingerfurche. Arch Klin Exp Dermatol 222:546–557.

Bass HN, Smith LE, Sparkes RS, Gyepes MT (1975): Case report 33. Syndrome Ident III(2):12–14.

Baughman FA (1971): CHANDS: The curly hair-ankyloblepharon-nail dysplasia syndrome. Birth Defects VII(8):100–102.

Beare JM (1952): Congenital pilar defect showing features of pili torti. Br J Dermatol 64:366–372.

Beçak W, Beçak ML, Andrade JD (1964): A genetical investigation of congenital analgesia. I. Cytogenetic studies. Acta Genet Stat Med 14:133–142.

Beckerman BL (1973): Lacrimal anomalies in anhidrotic ectodermal dysplasia. Am J Ophthalmol 75:728–730.

Beighton PH (1970): Familial hypertrichosis cubiti. J Med Genet 7:158–160.

Beighton P (1978): "Inherited Disorders of the Skeleton." Edinburgh: Churchill Livingston.

Beighton P, Hamersma H, Raad M (1979): Oculodento-osseous dysplasia: Heterogeneity or variable expression? Clin Genet 16:169–177.

Bellafiore V, Fiuffre I, Scialfa A (1966): La sindrome di Rothmund (Studio istologico cutaneo-lenticolare e citogenetico). Ann Ottalmol 92:394–407.

Berbich A, Dhermy P, Majbar M (1981): Ocular findings in a case of incontinentia pigmenti (Bloch-Sulzberger syndrome). Ophthalmologica 182:119–129.

Bergsma D (ed) (1979): "Birth Defects Compendium." 2nd ed. New York: Alan R Liss.

Berlin G (1961): Congenital generalized melanoleucoderma associated with hypodontia, hypotrichosis, stunted growth and mental retardation occurring in two brothers and two sisters. Dermatologica 123:227–243.

Bernard R, Giraud F, Touby M, Hartung M (1963): A propos de sept observations de dysplasie ectodermique chez des sujets de sexe feminin dont six dans la même famille. Discussion génétique. Arch Fr Pédiatr 20:1051–1061.

Bernirschke K, Lowry RB, Opitz JM, Schwarzacher HG, Spranger JW (1979): Developmental terms—Some proposals: First report on an International Working Group. Am J Med Genet 3:297–302.

Berres HH, Nitschke R (1968): Vergleichende klinische und morphologische Untersuchungen zwischen einem Neugeborenen mit Hypertrichosis universalis und gleichaltrigen hautgesunden Kindern. Z Kinderheilkd 102:327–340.

Bixler D, Poland C, Nancy WE (1973): Phenotypic variation in the popliteal pterygium syndrome. Clin Genet 4:220–228.

Bixler D, Spivack J, Bennett J, Christian JC (1972): The ectrodactyly-ectodermal dysplasia-clefting (EEC) syndrome. Clin Genet 3:43–51.

Blassingille B (1959): A rare form of ectodermal dysplasia—mental retardation, congenital skin disorder, and congenital spastic disorder. Neurology 9:384–391.

Blattner RJ (1968): Hereditary ectodermal dysplasia. J Pediatr 73:444–447.

Blinstrub RS, Lehman R, Sternberg TH (1964): Poikiloderma congenitale. Arch Dermatol 89:659–664.

Bollaert A, Wachholder A (1969): Dysplasie ectodermique. J Belge Radiol 52:201–209.

Böök JA (1950): Clinical and genetical studies of hypodontia. I. Premolar aplasia, hyperhidrosis, and canities prematura; a new hereditary syndrome in man. Am J Hum Genet 2:240–263.

Bopp C, Bernardi C (1974): Disqueratose congênita. Med Cutan Ibero-Lat Am 1:31–40.

Bowen P, Armstrong HB (1976): Ectodermal dysplasia, mental retardation, cleft lip/palate and other anomalies in three sibs. Clin Genet 9:35–42.

Bowen P, Armstrong HB (1979): Cleft lip-palate, ectodermal dysplasia and syndactyly. In Bergsma D (ed): "Birth Defects Compendium." New York: Alan R Liss.

Brauer A (1929): Hereditärer symmetrischer systematisierter Naevus aplasticus bei 38 Personen. Dermatol Wochenschr 89:1163–1168.

Brill CB, Hsu LYF, Hirschhorn K (1972): The syndrome of ectrodactyly, ectodermal dysplasia and cleft lip and palate: Report of a family demonstrating a dominant pattern. Clin Genet 3:295–302.

Brunoni D, Lederman H, Ferrari S, Farah L, Andrade J, Meloni VA (1982): Uma síndrome malformativa com nanismo mesomélico, malformações esqueléticas, displasia ectodérmica e facies típica. Ciênc Cult 34(suppl):694.

Brunoni D, Lederman H, Ferrari S, Farah LMS, Andrade JAD, Meloni VA (1984): Mesomelic dwarfism, skeletal abnormalities, ectodermal dysplasia. J Clin Dysmorph (in press).

Bryan HG, Nixon RK (1965): Dyskeratosis congenita and familial pancytopenia. JAMA 192:103–108.

Bühler EM (1982): Langer-Giedion syndrome and 8q- deletion. Am J Med Genet 11:359.

Buhr J von, Schuster B (1971): Zur Bedeutung der Frühdiagnose des Christ-Siemens-Touraine-Syndroms. Dtsch Ges Wesen 26:2230–2233.

Burket JM, Burket BJ, Burket DA (1983): Eyelid cysts, hypodontia, and hypotrichosis. J Am Acad Dermatol (in press).

Cantú JM, Arias J, Foncerrada M, Hernández A, Podoswa G, Rostenberg I, Macotela-Ruiz E (1975): Syndrome of onychotrichodysplasia with chronic neutropenia in an infant from consanguineous parents. Birth Defects XI(2):63–66.

Cantwell RJ (1975): Congenital sensori-neural deafness associated with onycho-osteo dystrophy and mental retardation (D.O.O.R. syndrome). Humangenetik 26:261–265.

Carey JC, Hall BD (1978): The Coffin-Siris syndrome. Five new cases including two siblings. Am J Dis Child 132:667–671.

Carney RG Jr (1976): Incontinentia pigmenti. Arch Dermatol 112:535–542.

Carney RG, Carney RG Jr (1970): Incontinentia pigmenti. Arch Dermatol 102:157–162.

Carvel RI (1969): Palmo-plantar hyperkeratosis and premature periodontal destruction. J Oral Med 24:73–82.

Cat I, Costa O, Freire-Maia N (1972): Odontotrichomelic hypohidrotic dysplasia. A clinical reappraisal. Hum Hered 22:91–95.

Cat I, Marinoni LP, Giraldi DJ, Furtado V de P, Pasquini R, Freire-Maia N, Braga H (1967): Epidermolysis bullosa dystrophica, hypoplastic type, associated with Pelger-Huet anomaly. J Med Genet 4:302–303.

Cat I, Marinoni LP, Moreira CA, Giraldi DJ, Costa O, Braga H (1971): Hipertricose lanuginosa universal. J Pediatr (Brazil) 36:26–28.

Chautard EA, Freire-Maia N (1970): Dermatoglyphic analysis in a highly mutilating syndrome. Acta Genet Med Gemellol 19:421–424.

Chautard-Freire-Maia EA, Primo-Parmo SL, Pinheiro M, Freire-Maia N (1981): Further evidence against linkage between Christ-Siemens-Touraine syndrome (CST) and XG loci. Hum Genet 57:205–206.

Clouston HR (1929): A hereditary ectodermal dystrophy. Can Med Assoc J 21:18–31.

Clouston HR (1939): The major forms of hereditary ectodermal dysplasia. Can Med Assoc J 40:1–7.

Coffin GS, Siris E (1970): Mental retardation with absent fifth fingernail and terminal phalanx. Am J Dis Child 119:433–439.

Coffin G, Siris E, Wegienka LC (1966): Mental retardation with osteocartilaginous anomalies. Am J Dis Child 112:205–213.

Cohen MM Jr (1974): The demise of the Marshall syndrome. J Pediatr 85:878.

Cohen MM Jr (1975): An etiologic and nosologic overview of craniosynostosis syndromes. Birth Defects XI(2):137–189.

Cohen MM Jr (1976): Dysmorphic syndromes with craniofacial manifestations. In Stewart RE, Prescott GH (eds): "Oral Facial Genetics." Saint Louis: C V Mosby.

Cohen MM Jr (1977): On the nature of syndrome delineation. Acta Genet Med Gemellol 26:103–119.

Cohen MM Jr (1978): Syndromes with cleft lip and cleft palate. Cleft Palate J 15:306–328.

Cohen MM Jr (1979): Ectrodactyly-ectodermal dysplasia-clefting syndrome. In Bergsma D (ed): "Birth Defects Compendium." 2nd ed. New York: Alan R Liss.

Cole HN, Giffen HK, Simmons JT, Stroud GM (1945): Congenital cataracts in sisters with congenital ectodermal dysplasia. JAMA 10:723–728.

Conterio F, Barrai I (1966): Effetti della consanguineità sulla mortalità e sulla morbilità nella popolazione della Diocesi di Parma. Att Ass Genet It 11:378–391.

Côté GB, Adamopoulos D, Pantelakis S (1982): Arthrogryposis and ectodermal dysplasia. Hum Hered 32:71–72.

Côté GB, Katsantoni A (1982): Osteosclerosis and ectodermal dysplasia. 3rd International Clinical Genetics Seminar, Athens.

Cram DL, Resneck JS, Jackson WB (1979): A congenital ichthyosiform syndrome with deafness and keratitis. Arch Dermatol 115:467–471.

Crawford JL (1970): Concomitant taurodontism and amelogenesis imperfecta in the American Caucasian. J Child Dent 37:171–175.

Crump IA, Danks DM (1971): Hypohidrotic ectodermal dysplasia. J Pediatr 78:466–473.

Curth HO, Warburton D (1965): The genetics of incontinentia pigmenti. Arch Dermatol 92:229–235.

Daentl DL, Frias JL, Gilbert EF, Opitz JM (1979): The Johanson-Blizzard syndrome: Case report and autopsy findings. Am J Med Genet 3:129–135.

Dahlberg PJ, Borer WZ, Newcomer KL, Yutuc WR (1983): Autosomal or X-linked recessive syndrome of congenital lymphedema, hypoparathyroidism, nephropathy, prolapsing mitral valve, and brachytelephalangy. Am J Med Genet 16:99–104.

Danforth CH (1925): Studies on hair with special reference to hypertrichosis. Arch Dermatol 12:380–401.

Danz DFG (1792): Sechste Bemerkung. Von Menschen ohne Haare und Zähne. Stark Arch Geburtsh Frauenz Neugeb Kinderkr 4:684.

Darwin C (1880): "De la Variation des Animaux et des Plantes à l'état Domestique." 2nd ed. Paris: C. Reinwald.

David RTM, Almeida FA, Araújo VC de, Marcucci G, Araújo NS de (1977): Lesões orais na síndrome de Jadassohn-Lewandowsky. An Bras Dermatol 52:235–242.

Davis JA (1966): A case of Swiss type agammaglobulinaemia and achondroplasia, demonstrated at the Royal Postgraduate Medical School. Br Med J 2:1371–1374.

Davison BCC (1965): Epidermolysis bullosa. J Med Genet 2:221–308.

Day DW, Israel JN (1978): Johanson-Blizzard syndrome. Birth Defects XIV(6B):275–287.

Dearborn GVN (1932): A case of congenital general pure analgesia. J Nerv Ment Dis 75:612–615.

De Boeck K, Degreef H, Verwilghen R, Corbee L, Casteels-Van Daele M (1981): Thrombocytopenia: First symptom in a patient with dyskeratosis congenita. Pediatrics 67:898–903.

Degos R, Bernard J, Belaïch S, Flandrin G, Varet B (1969): Syndrome de Zinsser-Cole-Engman (Zinsser-Fanconi). Bull Soc Fr Dermatol Syphiligr 76:15–17.

De Jager H (1965): Congenital anhidrotic ectodermal dysplasia. Case report. J Pathol Bacteriol 90:321–325.

Demis J, Weiner MA (1963): Alopecia universalis, onychodystrophy, and total vitiligo. Arch Dermatol 88:131–137.

Der Kaloustian VM, Kurban AK (1979): "Genetic Diseases of the Skin." Heidelberg: Springer-Verlag.

Desmons MMF, Bar J, Chevillard Y (1975): Érythrodermie ichthyosiforme congénitale sèche, surdi-mutité et hépatomégalie, de transmission récessive autosomique. Bull Soc Fr Dermatol Syphiligr 78:585–591.

Dinwiddie R, Gewitz M, Taylor JFN (1978): Cardiac defects in the Hallermann-Streiff syndrome. J Pediatr 92:77–78.

Dominok GW, Rönisch P (1968): Histologische Hautbefunde bei ektodermaler Dysplasie von anhydrotischen Typ. Dermatol Wochenschr 154:774–778.

Donnai D, Harris R (1979): A further case of a new syndrome including midface retraction, hypertrichosis, and skeletal anomalies. J Med Genet 16:483–486.

Durham RH (1960): "Encyclopedia of Medical Syndromes." New York: P.B. Hoeber.

Ellis RWB, van Creveld S (1940): A syndrome characterized by ectodermal dysplasia, polydactyly, chondro-dysplasia and congenital morbus cordis. Report of three cases. Arch Dis Child 15:65–84.

Epps DE, Mendonça BB, Olazabal LC, Billerbeck AEC, Wajntal A (1977): Poiquiloderma congênito familiar (S. de Rothmund-Thomson). Ciênc Cult 29(suppl):740.

Fadhil M, Ghabra TA, Deeb M, Der Kaloustian VM (1983): Odontoonychodermal dysplasia—A previously apparently undescribed ectodermal dysplasia. Am J Med Genet 14:335–346.

Falls HF, Schull WJ (1960): Hallermann-Streiff syndrome. A dyscephaly with congenital cataracts and hypotrichosis. Arch Ophthalmol 63:409–420.

Familusi JB, Jaiyesimi F, Ojo CO, Attah ED 'B (1975): Hereditary anhidrotic ectodermal dysplasia. Studies in a Nigerian family. Arch Dis Child 50:642–647.

Fára M (1971): Regional ectodermal dysplasia with total bilateral cleft. Acta Chir Plast 13:100–105.

Fára M, Gorlin RJ (1981): The question of hypertelorism in oculodentoosseous dysplasia. Am J Med Genet 10:101–102.

Feigin RD, Middelkamp JN, Kissane JM, Warren RJ (1971): Agammaglobulinemia and thymic dysplasia associated with ectodermal dysplasia. Pediatrics 47:143–147.

Feingold M (1978): The Coffin-Siris syndrome. Am J Dis Child 132:660–661.

Feinmesser M, Zelig S (1961): Congenital deafness associated with onychodystrophy. Arch Otolaryngol 74:507–508.

Felgenhauer W-R (1969): Hypertrichosis lanuginosa universalis. J Génét Hum 17:1–43.

Felix Rodriguez V, Ureta Huertos A, León LS de, Alonso Martin JA (1980): Displasia ectodermica anhidrótica parcial con acortamiento de miembro izquierdo. (Unpublished manuscript).

Felman AH, Frias JL (1977): The trichorhinophalangeal syndrome: Study of 16 patients in one family. Am J Roentgenol 129:631–638.

Felsher Z (1944): Hereditary ectodermal dysplasia. Arch Dermatol Syphilol 49:410–414.

Fischer E (1910): Ein Fall von erblicher Haararmut und die Art ihrer Vererbung. Ein Beitrag zur Familienanthropologie. Arch Rassenbiol Gesellschaftsbiol 7:50–56.

Fischer H (1921): Familiär hereditäres Vorkommen von Keratoma palmare et plantare, Nagelveränderungen, Haaranomalien und Verdickung der Endglieder der Finger und Zehen in 5 Generationen. (Die Beziehungen dieser Veränderungen zur inneren Sekretion.) Dermatol Zeitschr 32:114–142.

Fitch N (1980): The syndromes of Marshall and Weaver. J Med Genet 17:174–178.

Fonseca LG da, Freire-Maia N (1970): Congenital malformations of the limbs. Lancet i:90–91.

Franceschetti A (1953): Les dysplasies ectodermiques et les syndromes héréditaires apparentés. Dermatologica 106:129–156.

Franceschetti A, Jadassohn W (1954): A propos de l'"incontinentia pigmenti,"délimitation de deux syndromes différents figurant sous le même terme. Dermatologica 108:1–28.

Fraser GR, Friedmann AI (1967): "The Causes of Blindness in Childhood." Baltimore: The Johns Hopkins Press.

Freire-Maia A (1968): "Genética da Aquiropodia." PhD thesis, University of São Paulo, São Paulo, Brazil.

Freire-Maia A (1975): Genetics of acheiropodia (The handless and footless families of Brazil). VIII. Penetrance and expressivity. Clin Genet 7:98–102.

Freire-Maia A, Freire-Maia N, Morton NE, Azevêdo ES, Quelce-Salgado A (1975): Genetics of acheiropodia (The handless and footless families of Brazil). VI. Formal genetic analysis. Am J Hum Genet 27:521–527.

Freire-Maia A, Laredo-Filho J, Freire-Maia N (1978): Genetics of acheiropodia (The handless and footless families of Brazil). X. Roentgenologic study. Am J Med Genet 2:321–330.

Freire-Maia N (1969): Congenital skeletal limb deficiencies—a general view. Birth Defects V(3):7–13.

Freire-Maia N (1970a): A newly recognized genetic syndrome of tetramelic deficiencies, ectodermal dysplasia, deformed ears, and other abnormalities. Am J Hum Genet 22:370–377.

Freire-Maia N (1970b): Empirical risks in genetic counseling. Soc Biol 17:207–212.

Freire-Maia N (1971): Ectodermal dysplasias. Hum Hered 21:309–312.

Freire-Maia N (1973): Displasias ectodérmicas—Um grupo nosológico com contornos mal delineados. Acta Biol Parana 2:3–8.

Freire-Maia N (1975a): Some epidemiological and genetic aspects of congenital heart diseases. Acta Genet Med Gemellol 24:151–158.

Freire-Maia N (1975b): Adaptation and genetic load. In Salzano FM (ed): "The Role of Natural Selection in Human Evolution." Amsterdam: North-Holland.

Freire-Maia N (1975c): A heterozygote expression of a "recessive" gene? Hum Hered 25:302–304.

Freire-Maia N (1976): "Tópicos de Genética Humana." São Paulo: Hucitec and Edusp.

Freire-Maia N (1977a): Ectodermal dysplasias revisited. Acta Genet Med Gemellol 26:121–131.

Freire-Maia N (1977b): Nosologic groups. Hum Hered 27:251–256.

Freire-Maia N (1977c): Heterogeneity among ectodermal dysplasias. J Med Genet 14:234.

Freire-Maia N (1980): Alguns aspectos epidemiológicos e genéticos das cardiopatias congênitas. Arq Bras Cardiol 35:451–456.

Freire-Maia N (1982): Malformações, síndromes e displasias. Actas V Congr Latinoam Genét (Chile) 35–43.

Freire-Maia N (1984): Effects of consanguineous marriages on morbidity and precocious mortality: Genetic counseling. Am J Med Genet (in press).

Freire-Maia N, Azevedo JBC (1968): Skeletal limb deficiencies. Lancet ii:1296.

Freire-Maia N, Azevedo JBC (1977): Reduction deformities, twinning and mortality in Brazilian Whites and Negroes. Acta Genet Med Gemellol 26:133–140.

Freire-Maia N, Cat I, Costa O (1971): Ectodermal dysplasias. Excerpta Med, Int Congr Series no. 233. 4th Int Congr Hum Genet 68–69.

Freire-Maia N, Cat I, Lopes VLV, Chautard EA, Marçallo FA, Cavalli IJ, Pilotto RF, Schetino MC, Der Bedrossian AA (1970): A new malformation syndrome. Lancet i:840–841.

Freire-Maia N, Cat I, Rapone-Gaidzinski R (1977): An ectodermal dysplasia syndrome of alopecia, onychodysplasia, hypohidrosis, hyperkeratosis, deafness and other manifestations. Hum Hered 27:127–133.

Freire-Maia N, Felizali J, Figueiredo AC de, Opitz JM, Parreira M, Maia NA (1976): Hypertrichosis lanuginosa in a mother and son. Clin Genet 10:303–306.

Freire-Maia N, Fortes VA, Pereira LC, Opitz JM, Marçallo FA, Cavalli IJ (1975): A syndrome of hypohidrotic ectodermal dysplasia with normal teeth, peculiar facies, pigmentary disturbances, psychomotor and growth retardation, bilateral nuclear cataract, and other signs. J Med Genet 12:308–310.

Freire-Maia N, Freire-Maia A (1964): Multiple congenital abnormalities. Lancet i:113–114.

Freire-Maia N, Freire-Maia A (1967a): Malformations of the extremities. Lancet ii:367.

Freire-Maia N, Freire-Maia A (1967b): Recurrence risks of bone aplasias and hypoplasias of the extremities. Acta Genet 17:418–421.

Freire-Maia N, Guaraciaba MA, Quelce-Salgado A (1964): The genetical load in the Bauru Japanese isolate in Brazil. Ann Hum Genet 27:329–339.

Freire-Maia N, Laynes-de-Andrade F, Athayde-Neto A de, Cavalli IJ, Oliveira JC, Marçallo FA, Coelho A (1978): Genetic investigation in a northern Brazilian island. II. Random drift. Hum Hered 28:401–410.

Freire-Maia N, Pinheiro M (1979): Recessive anonychia totalis and dominant aplasia (or hypoplasia) of upper lateral incisors in the same kindred. J Med Genet 16:45–48.

Freire-Maia N, Pinheiro M (1980): So-called "anhidrotic ectodermal dysplasia." Int J Dermatol 19:455–456.

Freire-Maia N, Pinheiro M (1982a): Carrier detection in Christ-Siemens-Touraine syndrome (X-linked hypohidrotic ectodermal dysplasia). Am J Hum Genet 34:672–674.

Freire-Maia N, Pinheiro M (1982b): Ectodermal dysplasias in females. J Med Genet 19:316.

Freire-Maia N, Pinheiro M (1983a): Revisão clinicogenética de 99 displasias ectodérmicas (108 com inclusão das heterogeneidades). Ciênc Cult 35(suppl):694.

Freire-Maia N, Pinheiro M (1983b): Displasias ectodérmicas: Revisão clinico-genética. Ciênc Cult 35:577–579.

Freire-Maia N, Pinheiro M (1983c): Displasias ectodérmicas—uma revisão clinico-genética. An Bras Dermatol 58:213–214.

Freire-Maia N, Pinheiro M, Fernandes-dos-Santos A (1982): Xerodermia como sinal mais grave num quadro de displasia ectodérmica pura com etiologia genética. Ciênc Cult 34(suppl):764.

Freire-Maia N, Pinheiro M, Fernandes-dos-Santos A (1984): Trichoonychodysplasia with xerderma. An apparently hitherto undescribed pure ectodermal dysplasia. In preparation.

Freire-Maia N, Pinheiro M, Rapone-Gaidzinski R, Cat I (1981): A note on quantitation of intermammillary distance. Hum Hered 31:197–198.

Freire-Maia N, Quelce-Salgado A, Koehler RA (1959): Hereditary bone aplasias and hypoplasias of the upper extremities. Acta Genet Stat Med 9:33–40.

Freire-Maia N, Schetino MC, Der Bedrossian AA (1969): Dados clínicos e radiológicos sôbre um síndrome aparentemente novo. Ciênc Cult 21(suppl):280.

Frias JL, Smith DW (1968): Diminished sweat pores in hypohidrotic ectodermal dysplasia: A new method for assessment. J Pediatr 72:606–610.

Fried K (1972): Ectrodactyly-ectodermal dysplasia-clefting (EEC) syndrome. Clin Genet 3:396–400.

Fried K (1977): Autosomal recessive hydrotic ectodermal dysplasia. J Med Genet 14:137–139.

Fryns JP, Emmery L, Timmermans J, Pedersen JC, van der Berghe H (1980): Tricho-rhino-phalangeal syndrome type II: Langer-Giedion syndrome in a 2.5-year-old boy. J Génét Hum 28:53–56.

Fryns JP, Logghe N, van Eygen M, van der Berghe H (1981): Langer-Giedion syndrome and deletion of the long arm of chromosome 8. Hum Genet 58:231–232.

Fuks A, Rosenmann A, Chosack A (1978): Pseudoanodontia, cranial deformity, blindness, alopecia, and dwarfism: A new syndrome. J Dent Child 45:155–158.

Fukushima N, Anakura M, Arashima S, Matsuda I, Ohsawa T (1976): Tricho-rhino-phalangeal syndrome. The first case in Japan. Hum Genet 32:207–210.

Fulginiti VA, Hathaway WE, Pearlman DS, Kemp CH (1967): Agammaglobulinemia and achondroplasia. Br Med J 2:242.

Gagliardi ART, Prattesi R, Yazzle C, Varella MG, Tajara E (1983): Displasia ectodérmica associada a retardo de crescimento. Ciênc Cult 35(suppl):676.

Garb J (1958): Dyskeratosis congenita with pigmentation, dystrophia unguium, and leukoplakia oris. A follow-up report of two brothers. Arch Dermatol 77:704–712.

Gatti RA, Platt N, Pomerance HH, Hong R, Langer LO, Kay HEM, Good RA (1969): Hereditary lymphopenic agammaglobulinemia associated with a distinctive form of short-limbed dwarfism and ectodermal dysplasia. J Pediatr 75:675–684.

Gedde-Dahl T Jr (1971): "Epidermolysis Bullosa. A Clinical, Genetic and Epidemiological Study." Baltimore: Johns Hopkins Press.

Gellis SS, Feingold M (1979): Cranioectodermal dysplasia. Am J Med Genet 113:1275–1276.

Gemme G, Bonioli E, Ruffa G, Grosso P (1976): La sindrome EEC. Descrizione di due casi in una stessa famiglia. Minerva Pediatr 28:36–43.

German J (1969): Chromosome breakage syndromes. Birth Defects V(5):117–131.

German J (1972): Genes which increase chromosomal instability in somatic cells and predispose to cancer. Prog Med Genet VIII:61–101.

Giansanti JS, Hrabak RP, Waldron CA (1973): Palmar-plantar hyperkeratosis and concomitant periodontal destruction (Papillon-Lefèvre syndrome). Oral Surg 36:40–48.

Giansanti JS, Long SM, Rankin JL (1974): The "tooth and nail" type of autosomal dominant ectodermal dysplasia. Oral Surg 37:576–582.

Giedion A (1966): Das tricho-rhino-phalangeal Syndrom. Helv Paediatr Acta 21:475–482.

Giedion A, Burdea M, Fruchter Z, Meloni T, Trosc V (1973): Autosomal dominant transmission of the tricho-rhino-phalangeal syndrome. Report of 4 unrelated families, review of 60 cases. Helv Paediatr Acta 28:249–259.

Ginsburg LD, Sedano HO, Gorlin RJ (1970): Focal dermal hypoplasia syndrome. Am J Roentgenol 110:561–571.

Giraud F, Mattei JF, Rolland M, Ghiglione C, Santi PP de, Sudan N (1977): La dysplasie ectodermique de type Clouston. Arch Fr Pédiatr 34:982–993.

Goltz RW, Henderson RR, Hitch JM, Ott JE (1970): Focal dermal hypoplasia syndrome—A review of the literature and report of two cases. Arch Dermatol 101:1–11.

Goltz RW, Peterson WC, Gorlin RJ, Ravits HG (1962): Focal dermal hypoplasia. Arch Dermatol 86:707–711.

Goodman RM, Gorlin RJ (1977): "Atlas of the Face in Genetic Disorders." 2nd ed. Saint Louis: CV Mosby.

Goodman RM, Lockareff S, Gwinup G (1969): Hereditary congenital deafness with onychodystrophy. Arch Otolaryngol 90:474–477.

Gorlin RJ (1979): Oro-facio-digital syndrome I. In Bergsma D (ed): "Birth Defects Compendium." New York: Alan R Liss.

Gorlin RJ (1981): Lapsus-Caveat Emptor: Coffin-Lowry syndrome vs Coffin-Siris syndrome—An example of confusion compounded. Am J Med Genet 10:103–104.

Gorlin RJ (1982a): Thoughts on some new and old bone dysplasias. Skel Dyspl, Proc 3rd Int Clin Genet Semin, Athens, Greece. In Papadatos CJ, Bartsocas CS (eds): New York: Alan R Liss, pp 47–51.

Gorlin RJ, Červenka J (1975): Alopecia totalis, nail dysplasia and amelogenesis imperfecta. New Chromosomal and Malformation Syndromes. Birth Defects XI(5):23–24.

Gorlin RJ, Červenka J, Bloom BA, Langer LO Jr (1982): No chromosome deletion found on prometaphase banding in two cases of Langer-Giedion syndrome. Am J Med Genet 13:345–347.

Gorlin RJ, Chaudhry AP, Moss ML (1960): Craniofacial dysostosis, patent ductus arteriosus, hypertrichosis, hypoplasia of labia majora, dental and eye anomalies—a new syndrome? J Pediatr 56:778–785.

Gorlin RJ, Cohen MM Jr, Wolfson J (1969): Tricho-rhino-phalangeal syndrome. Am J Dis Child 118:595–599.

Gorlin RJ, Meskin LH, Peterson WC, Goltz RW (1963): Focal dermal hypoplasia syndrome. Acta Derm Venereol 43:421–440.

Gorlin RJ, Meskin LH, St Geme JW (1963): Oculodentodigital dysplasia. J Pediatr 63:69–75.

Gorlin RJ, Old T, Anderson VE (1970): Hypohidrotic ectodermal dysplasia in females. A critical analysis and argument for genetic heterogeneity. Z Kinderheilkd 108:1–11.

Gorlin RJ, Pindborg JJ (1964): "Syndromes of the Head and Neck." New York: McGraw-Hill.

Gorlin RJ, Pindborg JJ, Cohen MM Jr (1976): "Syndromes of the Head and Neck." New York: McGraw-Hill.

Gorlin RJ, Psaume J (1962): Orodigitofacial dysostosis—a new syndrome. A study of 22 cases. J Pediatr 61:520–530.

Gorlin RJ, Sedano H, Anderson VE (1964): The syndrome of palmar-plantar hyperkeratosis and premature periodontal destruction of the teeth. A clinical and genetic analysis of the Papillon-Lefèvre syndrome. J Pediatr 65:895–908.

Greene GW (1962): Genetic factors in ectodermal dysplasias. In Witkop CJ Jr (ed): "Genetics and Dental Health." New York: McGraw-Hill.

Gwinn JL, Lee FA (1974): Congenital anhidrotic ectodermal dysplasia. Am J Dis Child 128:215–216.

Hall JG, Pagon RA, Wilson KM (1980): Rothmund-Thomson syndrome with severe dwarfism. Am J Dis Child 134:165–169.

Haneke E (1979): The Papillon-Lefèvre syndrome: Keratosis palmoplantaris with periodontopathy. Hum Genet 51:1–35.

Harrod MJE, Stokes J, Peede LF, Goldstein JL (1976): Polycystic kidney disease in a patient with the oral-facial-digital syndrome—Type I. Clin Genet 9:183–186.

Hartwell SW, Pickrell K, Quinn G (1965): Congenital anhidrotic ectodermal dysplasia—Report of two cases. Clin Pediatr 4:383–386.

Hay RJ, Wells RS (1976): The syndrome of ankyloblepharon, ectodermal defects and cleft lip and palate: An autosomal dominant condition. Br J Dermatol 94:277–289.

Hazen PG, Zamora I, Bruner WE, Muir VA (1980): Premature cataracts in a family with hidrotic ectodermal dysplasia. Arch Dermatol 116:1385–1387.

Hecht F, Hecht BK, Austin WJ (1982): Incontinentia pigmenti in Arizona Indians including transmission from mother to son inconsistent with the half chromatid mutation model. Clin Genet 21:293–296.

Hernández A, Olivares F, Cantú JM (1979): Autosomal recessive onychotrichodysplasia, chronic neutropenia and mild mental retardation. Clin Genet 15:147–152.

Herrmann J (1979): Terminological, diagnostic, nosological, and anatomical-developmental aspects of developmental defects in man. Part II: Patient evaluation, delineation, and nosology of developmental defects—an overview. Adv Hum Genet 9:107–132.

Herrmann J, Gilbert EF, Opitz JM (1977): Dysplasia, malformations and cancer, especially with respect to the Wiedemann-Beckwith syndrome.

In Nichols WW, Murphy DG (eds): "Regularization of Cell Prolifera-
tion and Differentiation." New York: Plenum.

Herrmann J, Opitz JM (1974): Naming and nomenclature of syndromes.
Birth Defects X(7):69–86.

Holmes LB, Moser HW, Halldorsson S, Mack C, Pant SS, Matzilevich
B (1972): "Mental Retardation—An Atlas of Diseases with Associated
Physical Abnormalities." New York: MacMillan.

Hooft E, Jongbloet P (1964): Syndrome oro-digito-facial chez deux frères.
Arch Fr Pédiatr 21:729–740.

Howden EF, Oldenburg ThR (1972): Epidermolysis bullosa dystrophica:
Report of two cases. JADA 85:1113–1118.

Howell RR, Arbisser AI, Parsons DS, Scott CI, Fraustadt M, Collie
WR, Marshall RN, Ibarra OC (1981): The Sabinas syndrome. Am J
Hum Genet 33:957–967.

Hudson CD, Witkop CJ Jr (1975): Autosomal dominant hypodontia with
nail dysgenesis. Report of twenty-nine cases in six families. Oral Surg
39:409–423.

Hutt FB (1935): An earlier record of the toothless men of Sind. J Hered
26:65–66.

Iancu T, Komlos L, Shabtay F, Elian E, Halbrecht I, Böök JA (1975):
Incontinentia pigmenti. Clin Genet 7:103–110.

Ingle JI (1959): Papillon-Lefèvre syndrome: Precocious periodontosis with
associated epidermal lesions. J Periodontol 30:230–237.

Ishibashi A, Kurihara Y (1972): Golt's syndrome: Focal dermal dysplasia
syndrome (focal dermal hypoplasia). Dermatologica 144:156–167.

Jablonski S (1969): "Illustrated Dictionary of Eponymic Syndromes and
Diseases and their Synonyms." Philadelphia: WB Saunders.

Jackson ADM, Lawler SD (1951): Pachyonychia congenita: A report of
six cases in one family with a note on linkage data. Ann Eugen 16:141–
146.

Jacobsen AW (1928): Hereditary dystrophy of the hair and nails. JAMA
90:686–689.

Jansen LH (1951): The so-called "Dyskeratosis congenita." Dermatologica 103:167–177.

Jelinek JE, Bart RS, Schiff GM (1973): Hypomelanosis of Ito ("Incontinentia pigmenti achromians"). Report of three cases and review of the
literature. Arch Dermatol 107:596–601.

Jensen NE (1971): Congenital ectodermal dysplasia of the face. Br J
Dermatol 84:410–416.

Joachim H (1936): Hereditary dystrophy of the hair and nails in six
generations. Ann Int Med 10:400–402.

Joensen HD (1973): Epidermolysis bullosa dystrophica dominant in two families in the Faroe Islands. A clinico-genetic study of 56 living individuals. Acta Dermatol 53:53–60.

Johanson A, Blizzard R (1971): A syndrome of congenital aplasia of the alae nasi, deafness, hypothyroidism, dwarfism, absent permanent teeth, and malabsorption. J Pediatr 79:982–987.

Johnson VP, McMillin JM, Aceto Th Jr, Bruins G (1983): A newly recognized neuroectodermal syndrome of familial alopecia, anosmia, deafness and hypogonadism. Am J Med Genet 15:497–506.

Johnson VP, McMillin JM, Jaqua RJ (1982): A new syndrome of familial alopecia, anosmia, deafness, and hypogonadism. Birth Defects Conference, Birmingham, AL.

Jorgenson RJ (1971): Gingival fibromatosis. Birth Defects VII(7):278–280.

Jorgenson RJ (1974): Ectodermal dysplasia with hypotrichosis, hypohidrosis, defective teeth and unusual dermatoglyphics (Basan syndrome?). Birth Defects X(4):323–325.

Jorgenson RJ (1979): Oculo-mandibulo-facial syndrome. In Bergsma D (ed): "Birth Defects Compendium." 2nd ed. New York: Alan R Liss.

Jorgenson RJ, Pearlman A, Horton W (1975): Hallermann-Streiff syndrome. Birth Defects XI(2):391–393.

Jorgenson RJ, Warson RW (1973): Dental abnormalities in the trichodento-osseous syndrome. Oral Surg 36:693–700.

Joseph HL (1964): Pachyonychia congenita. Arch Dermatol 90:594–603.

Judge C, Chakanovskis JE (1971): The Hallermann-Streiff syndrome. J Ment Defic Res 15:115–120.

Kelley RI, Zackai EH, Charney EB (1982): Congenital hydronephrosis, skeletal dysplasia, and severe developmental retardation: The Schinzel-Giedion syndrome. J Pediatr 100:943–946.

Kelly TE, Wells HH, Tuck KB (1982): The Weissenbacher-Zweymüller syndrome: Possible neonatal expression of the Stickler syndrome. Am J Med Genet 11:113–119.

Kerr CB, Wells RS, Cooper KE (1966): Gene effect in carriers of anhidrotic ectodermal dysplasia. J Med Genet 3:169–176.

Kimura M (1959): Genetic load of a population and its significance in evolution. Jpn J Genet 35:7–33.

Kirkham TH, Werner EB (1975): The ophthalmic manifestations of Rothmund's syndrome. Can J Ophthalmol 10:1–14.

Kirman BH (1955): Idiocy and ectodermal dysplasia. Br J Dermatol 67:303–307.

Klein D (1950): Albinisme partiel (leucisme) avec surdi-mutité, blépharophimosis et dysplasie myo-ostéo-articulaire. Helv Paediatr Acta 5:38–58.

Klein D (1954): Cas observé. Une famille alsacienne de dysplasie ectodermique. J Génét Hum 3:210–213.

Kleinebrecht J, Degenhardt KH, Grubisic A, Günther E, Svejcar J (1981): Sweat pore counts in ectodermal dysplasias. Hum Genet 57:437–439.

Klingmüller G (1956): Über eigentümliche Konstitutionsanomalien bei 2 Schwestern und ihre Beziehungen zu neueren entwicklungspathologischen Befunden. Hautarzt 7:105–113.

Kolschütter A, Chappuis D, Meier C, Tönz O, Vassella F, Herschkowitz N (1974): Familial epilepsy and yellow teeth—a disease of the CNS associated with enamel hypoplasia. Helv Paediatr Acta 29:283–294.

Kopyść Z, Stańska M, Ryzko J, Kulczyk B (1980): The Saethre-Chotzen syndrome with partial bifid of the distal phalanges of the great toes. Hum Genet 56:195–204.

Koshiba H, Kimura O, Nakata M, Witkop CJ Jr (1978): Clinical, genetic, and histologic features of the trichoonychodental (TOD) syndrome. Oral Surg 46:376–385.

Kozlova SI, Altshuler BA, Kravchenko VL (1983): Self-limited autosomal recessive syndrome of skin ulceration, arthroosteolysis with pseudoacromegaly, keratitis, and oligodontia in a Kirghizian family. Am J Med Genet 15: 205–210.

Kratzsch R (1972): Ektodermale Dysplasie vom anhidrotischen Typ bei zwei Schwestern. Klin Pädiatr 184:328–332.

Kraus BS, Gottlieb MA, Meliton HR (1970): The dentition in Rothmund's syndrome. JADA 81:894–915.

Kristensen JK (1975): Poikiloderma congenitale—an early case of Rothmund-Thomson's syndrome. Acta Derm Venereol 55:316–318.

Kurwa AR, Abdel-Aziz A-HM (1973): Pili torti—congenital and acquired. Acta Derm Venereol 53:385–392.

Lamy ME (1969): Hereditary disorders of bones—an overview. Birth Defects V(4):8–16.

Laynes-de-Andrade F (1974): "Contribuição ao Estudo da Síndrome de Papillon-Lefèvre." Privat Dozent thesis, Federal University of Santa Catarina, Florianópolis, Brazil.

Leisti J, Sjöblom SM (1978): A new type of autosomal dominant trichodento-osseous syndrome. Proceedings of Birth Defects Conference (abstr.) XI:58. (Cf. Shapiro et al, 1983.)

Lelis II (1978): Autosomal recessive ectodermal dysplasia—A distinct nosological entity. Vestn Dermatol Venerol 12:56–59. (in Russian).

Lenz W (1963): "Medical Genetics." Chicago: Univ. Chicago Press.

Lerner AB (1961): Three unusual pigmentary syndromes. Arch Dermatol 83:151–159.

Levin LS, Perrin JCS, Ose L, Dorst JP, Miller JD, McKusick VA (1977): A heritable syndrome of craniosynostosis, short thin hair, dental abnormalities, and short limbs: Cranioectodermal dysplasia. J Pediatr 90:55–61.

Libis AS, Pinto IO, Viégas J (1982): Hipoplasia dérmica focal. Síndrome de Goltz. Med Cutan Ibero Lat Am 10:191–196.

Lichtenstein J, Warson R, Jorgenson R, Dorst JP, McKusick VA (1972): The tricho-dento-osseous (TDO) syndrome. Am J Hum Genet 24:569–582.

Lodin H, Sjögren I (1964): Chondro-ectodermal dysplasia (Ellis-van Creveld's syndrome). Two certain and two probable cases in the same family. Acta Paediatr 53:583–590.

Lowry RB, Robinson GC, Miller JR (1966): Hereditary ectodermal dysplasia. Symptoms, inheritance patterns, differential diagnosis, management. Clin Pediatr 5:395–402.

Machtens E, von Weyhrother H-G, Brands Th, Marxhors R (1972): Klinische Aspekte der ektodermalen Dysplasie. Z Kinderheilkd 112:265–280.

Majewski F, Spranger J (1976): Case report 49. Syndrome Ident. IV(2):17–21.

Marçallo FA, Freire-Maia N, Azevedo JBC, Simões IA (1964): Inbreeding effect on mortality in South Brazilian populations. Ann Hum Genet 27:203–217.

Mardini MK, Ghandour M, Sakati NA, Nyhan WL (1978): Johanson-Blizzard syndrome in a large inbred kindred with three involved members. Clin Genet 14:247–250.

Maroteaux P (1969): Spondyloepiphyseal dysplasias and metatropic dwarfism. Birth Defects V(4):35–47.

Marquina A, Aliaga A, Fortea JM, Oliver V, Segarra M (1975): Síndrome de Rothmund-Thomson. Actas Dermosifiliogr 66:59–64.

Marshall D (1958): Ectodermal dysplasia. Report of kindred with ocular abnormalities and hearing defect. Am J Ophthalmol 45:143–156.

Marshall RE, Graham CB, Scott CR, Smith DW (1971): Syndrome of accelerated skeletal maturation and relative failure to thrive: A newly recognized clinical growth disorder. J Pediatr 78:95–101.

Mattei JF, Laframboise R, Rouault F, Giraud F (1981): Coffin-Lowry syndrome in sibs. Am J Med Genet 8:315–319.

McKenzie J (1958): The first arch syndrome. Arch Dis Child 33:477–486.

McKenzie J (1968): The first arch syndrome and associated anomalies. In Longacre JJ (ed): "Craniofacial Anomalies: Pathogenesis and Repair." Philadelphia: Lippincott.

McKusick VA (1975): "Mendelian Inheritance in Man. Catalogs of Autosomal Dominant, Autosomal Recessive and X-Linked Phenotypes." 4th ed. Baltimore: Johns Hopkins University Press.

McKusick VA (1979): "Mendelian Inheritance in Man. Catalogs of Autosomal Dominant, Autosomal Recessive and X-Linked Phenotypes." 5th ed. Baltimore: Johns Hopkins University Press.

McKusick VA (1983): "Mendelian Inheritance in Man. Catalogs of Autosomal Dominant, Autosomal Recessive and X-Linked Phenotypes." 6th ed. Baltimore: Johns Hopkins University Press.

McKusick VA, Cross HE (1966): Ataxia-telangiectasia and Swiss-type agammaglobulinemia. Two genetic disorders of the immune mechanism in related Amish sibships. JAMA 195:739–745.

McKusick VA, Egeland JA, Eldridge R, Krusen DE (1964): Dwarfism in the Amish. I. The Ellis-van Creveld syndrome. Bull Johns Hopkins Hosp 115:306–336.

McNaughton PZ, Pierson DL, Rodman OG (1976): Hidrotic ectodermal dysplasia in a black mother and daughter. Arch Dermatol 112:1448–1450.

Menkes JH, Alter M, Steigleder GK, Weakley DR, Sung JH (1962): A sex-linked recessive disorder with retardation of growth, peculiar hair, and focal cerebral and cerebellar degeneration. Pediatrics 29:764–779.

Messow K, Götz A, Murken JD, Rodewald A, Riegel K (1977): Ektodermale Dysplasie vom anhidrotischen Typ—Identifizierung heterozygoter Merkmalsträgerinnen. Z Geburtsch Perinatol 181:129–133.

Mikaelian DO, Der Kaloustian VM, Shahin NA, Barsoumian VM (1970): Congenital ectodermal dysplasia with hearing loss. Arch Otolaryngol 92:85–89.

Mochizuki Y, Teramura F, Kawai K (1971): A family of congenital anhidrotic ectodermal dysplasia. Ann Paediatr Jpn 17:31–39.

Moghadam H, Statten P (1972): Hereditary sensorineural hearing loss associated with onychodystrophy and digital malformations. Can Med Assoc J 107:310–312.

Montgomery H (1967): "Dermatopathology." New York: Harper and Row.

Morgan JD (1971): Incontinentia pigmenti (Bloch-Sulzberger syndrome). Am J Dis Child 122:294–300.

Morris J, Ackerman AB, Koblenzer PJ (1969): Generalized spiny hyperkeratosis, universal alopecia and deafness. A previously undescribed syndrome. Arch Dermatol 100:693–698.

Morton NE, Crow JF, Muller HJ (1956): An estimate of the mutational damage in man from data on consanguineous marriages. Proc Natl Acad Sci (USA) 42:855–863.

Moynahan EJ (1970): XTE syndrome (xeroderma, talipes and enamel defect): A new heredo-familial syndrome. Proc R Soc Med Lond 63:1–2.

Munford AG (1976): Papillon-Lefèvre syndrome: Report of two cases in the same family. JADA 93:121–124.

Murachi S, Nogami H, Oki T, Ogino T (1981): Familial tricho-rhino-phalangeal syndrome type II. Clin Genet 19:149–155.

Murray FA (1921): Congenital anomalies of the nails. Four cases of hereditary hypertrophy of nail-bed associated with a history of erupted teeth at birth. Br J Dermatol 33:409–411.

Myers EN, Stool SE, Koblenzer PJ (1971): Congenital deafness, spiny hyperkeratosis, and universal alopecia. Arch Otolaryngol 93:68–74.

Naegeli O (1927): Familiärer Chromatophorennävus. Schweiz Med Wochenschr 57:48–49.

Nakata M, Koshiba H, Eto K, Nance WE (1980): A genetic study of anodontia in X-linked hypohidrotic ectodermal dysplasia. Am J Hum Genet 32:908–919.

Nazzaro P, Argentieri R, Bassetti F, Leonetti F, Topi G, Valenzano L (1972): Dyskeratose congénitale de Zinsser-Cole-Engmann. Bull Soc Fr Dermatol Syphiligr 79:242–244.

Nelson WE, Vaughan VC, McKay RJ (1969): "Textbook of Pediatrics." Philadelphia: WB Saunders.

O'Donnell JJ, Sirkin S, Hall BD (1976): Generalized osseous abnormalities in the Marshall syndrome. Birth Defects XII(5):299–314.

Olinsky A, Thomson PD (1970): Anhidrotic ectodermal dysplasia: Presentation in the neonatal period. S Afr Med J 44:1234–1235.

Opitz JM (1979): Terminological, diagnostic, nosological, and anatomical-developmental aspects of developmental defects in man. Part I: Terminological and epistemological considerations of human malformations. Adv Hum Genet 9:71–107.

Opitz JM (1981a): The developmental field concept in clinical genetics. Symposium at the Lake Wilderness Continuing Education Center of the University of Washington. (Unpublished manuscript.)

Opitz JM (1981b): "Recent Topics in Clinical Genetics." Helena: Shodair Children's Hospital.

Opitz JM, Gilbert EF (1981): Pathogenetic analysis of congenital anomalies in humans. (Unpublished manuscript.)

Opitz JM, Herrmann J, Dieker H (1969): The study of malformation syndromes in man. Birth Defects V(2):1–10.

O'Rourk TR Jr, Bravos A (1969): An oculo-dento-digital dysplasia. Birth Defects V(2):226–227.

Pabst H, Groth O (1975): Unusual ectodermal dysplasia. Lancet ii:824.

Pabst HF, Groth O, McCoy EE (1981): Hypohidrotic ectodermal dysplasia with hypothyroidism. J Pediatr 98:223–227.

Pantke OA, Cohen MM Jr, Witkop CJ Jr, Feingold M, Schaumann B, Pantke HC, Gorlin RJ (1975): The Saethre-Chotzen syndrome. Birth Defects XI(2):190–225.

Passarge E (1965): Epidermolysis bullosa hereditaria simplex. A kindred affected in four generations. J Pediatr 67:819–825.

Passarge E, Fries E (1973): X chromosome inactivation in X-linked hypohidrotic ectodermal dysplasia. Nature 245:58–59.

Passarge E, Fries E (1977): Autosomal recessive hypohidrotic ectodermal dysplasia with subclinical manifestation in the heterozygote. Birth Defects XIII(3C):95–100.

Passarge E, Nuzum CT, Schubert WK (1966): Anhidrotic ectodermal dysplasia as autosomal recessive trait in an inbred kindred. Humangenetik 3:181–185.

Peltola J, Kuokkanen K (1978): Tricho-rhino-phalangeal syndrome in five successive generations: Report of a family in Finland. Acta Dermatol 58:65–68.

Perabo F, Velasco JA, Prader A (1956): Ektodermale Dysplasie vom anhidrotischem Typus. 5 neue Beobachtungen. Helv Paediatr Acta 11:604–639.

Perlman MM (1971): Incontinentia pigmenti associated with raised alkaline phosphatase and disturbance of the plasma proteins. S Afr Med J 45:1334–1336.

Peterson-Falzone SJ, Caldarelli DD, Landahl KL (1981): Abnormal laryngeal vocal quality in ectodermal dysplasia. Arch Otolaryngol 107:300–304.

Pfeiffer RA (1960): Zur Frage der Vererbung der Incontinentia pigmenti Bloch-Siemens. Z menschl Vererb Konstitutionsl 35:469–493.

Picarelli A (1968): Le parodontopatie giovanili in corso d'ipercheratosi palmo-plantare e la sindrome di Papillon-Lefèvre. Minerva Stomatol 17:587–601.

Piguet B, Coumel C, Bonnefond-Craponne M (1969): Kératodermie palmo-plantaire (type Papillon-Lefèvre). Rev Stomatol Chir Maxillofac 70:446–459.

Pilotto RF: "Estudo Genético-Clínico da Síndrome de Ellis-van Creveld." MSc thesis, Federal University of Paraná, Curitiba, Brazil.

Pilotto RF, Petrelli NE, Marçallo FA, Pacheco CNA, Parolin BA (1976): Síndrome tricorrinofalangeano. Relato de um caso. Ciênc Cult 28(suppl):307.

Pinheiro M (1977): "Síndrome de Christ-Siemens-Touraine—Estudos Genéticos e Clinicos de 40 Casos em uma Familia." MSc thesis, Federal University of Paraná, Curitiba, Brazil.

Pinheiro M (1983): "Displasias Ectodérmicas do grupo A—Classificacão, Etiologia Genética e Descrição de Duas Afecções Novas." PhD thesis, University of São Paulo, São Paulo, Brazil.

Pinheiro M, Freire-Maia N (1977): Identifying carriers for X-linked hypohidrotic ectodermal dysplasia. Lancet ii:936.

Pinheiro M, Freire-Maia N (1979a): Christ-Siemens-Touraine syndrome—A clinical and genetic analysis of a large Brazilian kindred. I. Affected females. Am J Med Genet 4:113–122.

Pinheiro M, Freire-Maia N (1979b): Christ-Siemens-Touraine syndrome—A clinical and genetic analysis of a large Brazilian kindred. II. Affected males. Am J Med Genet 4:123–128.

Pinheiro M, Freire-Maia N (1979c): Christ-Siemens-Touraine syndrome—A clinical and genetic analysis of a large Brazilian kindred. III. Carrier detection. Am J Med Genet 4:129–134.

Pinheiro M, Freire-Maia N (1980): EEC and odontotrichomelic syndromes. Clin Genet 17:363–364.

Pinheiro M, Freire-Maia N (1981a): Tricho-odonto-onicodisplasia: Quatro casos em uma irmandade. Ciênc Cult 33(suppl):696.

Pinheiro M, Freire-Maia N (1981b): Odonto-onicodisplasia com alopecia: Dois casos em uma irmandade. Ciênc Cult 33(suppl):696.

Pinheiro M, Freire-Maia N (1982): Uma displasia ectodérmica pura devida a gene autossômico dominante. Ciênc Cult 34(suppl):764.

Pinheiro M, Freire-Maia N (1983): Dermoodontodysplasia: An eleven-member, four generation pedigree with an apparently hitherto undescribed pure ectodermal dysplasia. Clin Genet 24:58–68.

Pinheiro M, Freire-Maia N, Chautard-Freire-Maia EA (1982): Displasia ectodérmica e diabete lipoatrófica em uma paciente filha de casamento consangüíneo. Ciênc Cult 34(suppl):764.

Pinheiro M, Freire-Maia N, Chautard-Freire-Maia EA, Araújo LMB, Liberman B (1983a): Uma síndrome acro-renal com sinais de displasia ectodérmica, malformações ósseas, diabete lipoatrófica e outras manifestações. Ciênc Cult 35(suppl):694.

Pinheiro M, Freire-Maia N, Chautard-Freire-Maia EA, Araújo LMB, Liberman B (1983b): Aredyld: A syndrome combining an acrorenal field defect, ectodermal dysplasia, lipoatrophic diabetes, and other manifestations. Am J Med Genet 16:29–33.

Pinheiro M, Freire-Maia N, Gollop TR (1984): Odontoonychodysplasia with alopecia—a new pure ectodermal dysplasia with probable autosomal recessive inheritance. Am J Med Genet (in press).

Pinheiro M, Freire-Maia N, Roth AJ (1983c): Trichoodontoonychial dysplasia—A new meso-ectodermal dysplasia. Am J Med Genet 15:67–70. (See also Am J Med Genet 16:143–144, 1983).

Pinheiro M, Ideriha MT, Chautard-Freire-Maia EA, Freire-Maia N, Primo-Parmo SL (1981a): Christ-Siemens-Touraine syndrome. Investigations on two large Brazilian kindreds with a new estimate of the manifestation rate among carriers. Hum Genet 57:428–431.

Pinheiro M, Pereira LC, Freire-Maia N (1980): Síndrome triconicodontósseo—uma nova entidade com displasia ectodérmica e malformações. Ciênc Cult 32 (suppl):691.

Pinheiro M, Pereira LC, Freire-Maia N (1981b): A previously undescribed condition: Tricho-odonto-onycho-dermal syndrome. A review of the tricho-odonto-onychial subgroup of ectodermal dysplasias. Br J Dermatol 105:371–382.

Pinsky L (1974): A community of human malformation syndromes involving the Müllerian ducts, distal extremities, urinary tract, and ears. Teratology 9:65–80.

Pinsky L (1975): The community of human malformation syndromes that shares ectodermal dysplasia and deformities of the hands and feet. Teratology 11:227–242.

Pinsky L (1977): The polythetic (phenotypic community) system of classifying human malformation syndromes. Birth Defects XIII(3A):13–30.

Pinsky L, DiGeorge AM (1966): Congenital familial sensory neuropathy with anhidrosis. J Pediatr 68:1–13.

Piussan C, Carton F, Risbourg B, Guerrier C (1973): L'incontinentia pigmenti. Ses aspects cliniques, histologiques et ultra-structuraux. Ann Pédiatr 20:223–238.

Podoswa-Martinez G, Laguna Ocampo O, Armendares SS (1973): Alopecia universal congenita asociada con otros trastornos del ectodermo—Estudio y presentación de tres casos. (Unpublished manuscript.)

Preus M, Fraser FC (1973): The lobster claw defect with ectodermal defects, cleft lip-palate, tear duct anomaly and renal anomalies. Clin Genet 4:369–375.

Pries C, Mittelman D, Miller M, Solomon LM, Pashayan HM, Pruzansky S (1974): The EEC syndrome. Am J Dis Child 127:840–844.

Pruzansky S, Pashayan H, Kreiborg S, Miller M (1975): Roentgencephalometric studies of the premature craniofacial synostoses: Report of a family with the Saethre-Chotzen syndrome. Birth Defects XI(2):226–237.

Qazi QH, Smithwick EM (1970): Triphalangy of thumbs and great toes. Am J Dis Child 120:255–257.

Rajagopalan K, Tay CH (1977): Hidrotic ectodermal dysplasia. Study of a large Chinese pedigree. Arch Dermatol 113:481–485.

Rapone-Gaidzinski R (1978): "Displasias Ectodérmicas—Revisão Clínico-Genética com Especial Referência ao Problema da Sudorese." MSc thesis, Federal University of Paraná, Curitiba, Brazil.

Rapp RS, Hodgkin WE (1968): Anhidrotic ectodermal dysplasia: Autosomal dominant inheritance with palate and lip anomalies. J Med Genet 5:269–272.

Reddy BSN, Chandra S, Singh G (1978): Anhidrotic ectodermal dysplasia. Int J Dermatol 17:139–141.

Reed T, Schreiner RL (1983): Absence of dermal ridge patterns: Genetic heterogeneity. Am J Med Genet 16:81–88.

Reisner SH, Kott E, Bornstein B, Salinger H, Kaplan I, Gorlin RJ (1969): Oculodentodigital dysplasia. Am J Dis Child 118:600–607.

Rimoin DL, Edgerton MT (1967): Genetic and clinical heterogeneity in the oral-facial-digital syndromes. J Pediatr 71:94–102.

Rimoin DL, Hall J, Maroteaux P (1979): International nomenclature of constitutional diseases of bone with bibliography. Birth Defects XV(10).

Robinson GC, Johnston MM (1967): Pili torti and sensory neural hearing loss. J Pediatr 70:621–623.

Robinson GC, Miller JR, Bensimon JR (1962): Familial ectodermal dysplasia with sensori-neural deafness and other anomalies. Pediatrics 30:797–802.

Robinson GC, Miller JR, Worth HM (1966): Hereditary enamel hypoplasia: Its association with characteristic hair structure. Pediatrics 37:498–502.

Robinson GC, Wildervanck LS, Chiang TP (1973): Ectrodactyly, ectodermal dysplasia, and cleft lip-palate syndrome. Its association with conductive hearing loss. J Pediatr 82:107–109.

Ronchese F (1932): Twisted hairs (pili torti). Arch Dermatol Syphilol 26:98–109.

Rosemberg S, Carneiro PC, Zerbini MCN, Gonzalez CH (1983): Chondroectodermal dysplasia (Ellis-van Creveld) with anomalies of CNS and urinary tract. Am J Med Genet 15:291–295.

Rosenmann A, Shapira T, Cohen MM Jr (1976): Ectrodactyly, ectodermal dysplasia and cleft palate (EEC syndrome). Clin Genet 9:347–353.

Rosselli D, Gulienetti R (1961): Ectodermal dysplasia. Br J Plast Surg 14:190–204.

Rousset MJ (1952): Génodermatose difficilement classable (Trichorrhexis nodosa) prédominant chez les mâles dans quatre générations. Bull Soc Fr Dermatol Syphiligr 59:298–300.

Rubin A (1967): "Handbook of Congenital Malformations." Philadelphia: WB Saunders.

Rubin MB (1972): Incontinentia pigmenti achromians. Multiple cases within a family. Arch Dermatol 105:424–425.

Rüdiger RA, Haase W, Passarge E (1970): Association of ectrodactyly, ectodermal dysplasia, and cleft lip-palate. Am J Dis Child 120:160–163.

Ruiz-Maldonado R, Carnevale A, Tamayo L, Montiel EM (1974): Focal dermal hypoplasia. Clin Genet 6:36–45.

Rycroft RJG, Moynahan EJ, Wells RS (1976): Atypical ichthyosiform erythroderma, deafness and keratitis. Br J Dermatol 94:211–217.

Šalamon T, Cubela V, Bogdanović B, Lazović O, Bulatović N (1967): Über ein Geschwisterpaar mit einer eigenartigen ektodermalen Dysplasie. Arch Klin Exp Dermatol 230:60–68.

Šalamon T, Lazović O, Stenek S (1972): Über Haarveränderungen bei einer mit multiplen Anomalien einhergehenden ektodermalen Dysplasie. Hautarzt 23:441–445.

Šalamon T, Maherle G, Beighton E (1974): The hairs in the syndrome of Netherton and in a peculiar form of ectodermal dysplasia. Dermatol Monatsschr 160:362–372.

Šalamon T, Miličevič M (1964): Über eine besondere Form der ektodermalen Dysplasie mit Hypohidrosis, Hypotrichosis, Hornhautveränderungen, Nagel- und anderen Anomalien bein einem Geschwisterpaar. Arch Klin Exp Dermatol 220:564–575.

Saldanha PH, Schmidt BJ, Leon N (1964): A genetical investigation of congenital analgesia. II. Clinico-genetical studies. Acta Genet Stat Med 14:143–158.

Salinas CF, Spector M (1979): Tricho-dental syndrome. A new syndrome with autosomal dominant inheritance. J Dent Res 58:268.

Salinas CF, Spector M (1980): Tricho-dental syndrome. In Brown AC, Crounse RG (eds): "Hair, Trace Elements, and Human Illness." New York: Praeger,

Salmon MA, Lindenbaum RH (1978): "Developmental Defects and Syndromes." Aylesbury: HM+M.

Salzano FM, Marçallo FA, Freire-Maia N, Krieger H (1962): Genetic load in Brazilian Indians. Acta Genet Stat Med 12:212–218.

Samuelson G (1970): Hereditary ectodermal dysplasia. Report of two cases. Acta Paediatr Scand 59:94–99.

Sánchez O, Hazas JJM, DeMatos IO de (1981): The deafness, onycho-osteo-dystrophy, mental retardation syndrome. Hum Genet 58:228–230.

Schinzel A (1980): A case of multiple skeletal anomalies, ectodermal dysplasia, and severe growth and mental retardation. Helv Paediatr Acta 35:243–251.

Schinzel A, Giedion A (1978): A syndrome of severe midface retraction, multiple skull anomalies, clubfeet, and cardiac and renal malformations in sibs. Am J Med Genet 1:361–375.

Schnitzler A, Schubert B, Larget-Piet L, Berthelot J, Cleirens S, Taviaux D (1978): Le syndrome de Rudiger (syndrome EEC). A propos d'un cas associé à un eczéma atopique. Ann Dermatol Venerol 105:201–206.

Schnyder UW (1967): New findings in the ichthyosis and epidermolysis group. Proc 3rd Int Congr Hum Genet, Chicago (1966). Baltimore: Johns Hopkins University Press.

Schönfeld PHIR (1980): The pachyonychia congenita syndrome. Acta Dermatol Venereol 60:45–49.

Schöpf E, Schulz H-J, Passarge E (1971): Syndrome of cystic eyelids, palmo-plantar keratosis, hypodontia and hypotrichosis as a possible autosomal recessive trait. Birth Defects XII(8):219–221.

Scoggins RB, Prescott KJ, Asher GH, Blaylock WK, Bright RW (1971): Dyskeratosis congenita with Fanconi-type anemia: Investigations of immunologic and other defects. Clin Res 19:409.

Senior B (1971): Impaired growth and onychodysplasia—Short children with tiny toenails. Am J Dis Child 122:7–9.

Sensenbrenner JA, Dorst JP, Owens RP (1975): New syndrome of skeletal, dental and hair anomalies. Birth Defects XI(2):372–379.

Senter TP, Jones KL, Sakati N, Nyhan WL (1978): Atypical ichthyosiform erythroderma and congenital neurosensory deafness—a distinct syndrome. J Pediatr 92:68–72.

Setleis H, Kramer B, Valcarcel M, Einhorn AH (1963): Congenital ectodermal dysplasia of the face. Pediatrics 32:540–548.

Settineri WMF (1974): "Estudos Genéticos e Clínicos em uma Forma de Displasia Ectodérmica." MSc thesis, Federal University of Rio Grande do Sul, Porto Alegre, Brazil.

Settineri WMF, Salzano FM, Melo e Freitas MJ de (1976): X-linked anhidrotic ectodermal dysplasia with some unusual features. J Med Genet 13:212–216.

Shapira Y, Yatziv S, Deckelbaum R (1982): Growth retardation, alopecia, pseudoanodontia and optic atrophy. Syndr Ident 8:14–16.

Shapiro SD, Quattromani FL, Jorgenson RJ, Young RS (1983): Trichodento-osseous syndrome: Heterogeneity or clinical variability. Am J Med Genet 16:225–236.

Shore SW (1970): Ectodermal dysplasia: A case report. J Dent Child 37:254–257.

Silengo MC, Davi GF, Bianco R, Costa M, DeMarco A, Verona R, Franceschini P (1982): Distinctive hair changes (pili torti) in Rapp-Hodgkin ectodermal dysplasia syndrome. Clin Genet 21:297–300.

Silva EO da, Janovitz D, Albuquerque SC (1980) Ellis-van Creveld syndrome: Report of 15 cases in an inbred kindred. J Med Genet 17:349–356.

Sirinavin C, Trowbridge AA (1975): Dyskeratosis congenita: Clinical features and genetic aspects. J Med Genet 12:339–354.

Skinner BA, Greist MC, Norins AL (1981): The keratitis, ichthyosis, and deafness (KID) syndrome. Arch Dermatol 117:285–289.

Smith DS (1974): Comment. J Pediatr 84:553.

Smith DW (1969): Recognizable patterns of malformations in childhood. Birth Defects V(2):255–272.

Smith DW (1970): "Recognizable Patterns of Human Malformation. Genetic, Embryologic and Clinical Aspects." Philadelphia: WB Saunders.

Smith DW (1975): Classification, nomenclature, and naming of morphologic defects. J Pediatr 87:162–164.

Smith DW (1981): "Recognizable Patterns of Human Deformation." Philadelphia: WB Saunders.

Smith DW (1982): "Recognizable Patterns of Human Malformation. Genetic, Embryologic and Clinical Aspects." 3rd ed. Philadelphia: WB Saunders.

Smith DW, Knudson RW (1977): Aberrant scalp hair patterning in hypohidrotic ectodermal dysplasia. J Pediatr 90:248–250.

Snyder CH (1965): Syndrome of gingival hyperplasia, hirsutism, and convulsions. J Pediatr 67:499–502.

Sofaer JA (1981a): A dental approach to carrier screening in X-linked hypohidrotic ectodermal dysplasia. J Med Genet 18:459–460.

Sofaer JA (1981b): Hypodontia and sweat pore counts in detecting carriers of X-linked hypohidrotic ectodermal dysplasia. Br Dent J 151:327–330.

Solomon LM, Esterly NB (1973): "Neonatal Dermatology." Philadelphia: WB Saunders.

Solomon LM, Keuer J (1980): The ectodermal dysplasias. Problems of classification and some newer syndromes. Arch Dermatol 116:1295–1299.

Sorrow JR Jr, Hill C, Hitch JM (1963): Dyskeratosis congenita. First report of its occurrence in a female and a review of the literature. Arch Dermatol 88:156–162.

Spranger J, Benirschke K, Hall JG, Lenz W, Lowry RB, Opitz JM, Pinsky L, Schwarzacher HG, Smith DW (1982): Errors of morphogenesis: Concepts and terms. Recommendations of an International Working Group. J Pediatr 100:160–165.

Sri-Skanda-Rajah-Sivayoham I, Ratnaike VT (1975): Rothmund-Thomson syndrome in an Oriental patient. Ann Ophthalmol 7:417–420.

Stasiowska B, Sartoris, S, Goitre M, Benso L (1981): Rapp-Hodgkin ectodermal dysplasia syndrome. Arch Dis Child 56:793–796.

Steele RW, Bass JW (1970): Hallermann-Streiff syndrome. Clinical and prognostic considerations. Am J Dis Child 120:462–465.

Steier W, van Voolen GA, Selmanowitz VJ (1972): Dyskeratosis congenita: Relationship to Fanconi's anemia. Blood 39:510–521.

Stevanović DV (1959): Alopecia congenita. The incomplete dominant form of inheritance with varying expressivity. Acta Genet Stat Med 9:127–132.

Stewart RE, Prescott GH (eds) (1976): "Oral Facial Genetics." Saint Louis: CV Mosby.

Stieglitz JB, Centerwall WR (1983): Pachyonychia congenita (Jadassohn-Lewandowsky syndrome): A seventeen-member, four-generation pedigree with unusual respiratory and dental involvement. Am J Med Genet 14:21–28.

Strandberg J (1922): A contribution to our knowledge of aplasia moniliformis. Acta Derm Venereol 3:650–655.

Streiff EB (1950): Dysmorphie mandibulo-faciale (tête d'oiseau) et altérations oculaires. Ophthalmologica 120:79–83.

Sugarman GI, Katakia M, Menkes J (1971): See-saw winking in a familial oral-facial-digital syndrome. Clin Genet 2:248–254.

Summitt RL, Hiatt RL (1971): Hypohidrotic ectodermal dysplasia with multiple associate anomalies. Birth Defects VII(8):121–124.

Suskind R, Esterly NB (1971): Congenital hypertrichosis universalis. Birth Defects VII(8):103–106.

Swallow JN, Gray OP, Harper PS (1973): Ectrodactyly, ectodermal dysplasia and cleft lip and palate (EEC syndrome). Br Assoc Dermatol. Fifty-third annual meeting. 12.1–12.3.

Swanson AG (1963): Congenital insensitivity to pain with anhydrosis. Arch Neurol 8:299–306.

Takematsu H, Sato S, Igarashi M, Seiji M (1983): Incontinentia pigmenti achromians (Ito). Arch Dermatol 119:391–395.

Taylor NB (1957b): "Stedman's Medical Dictionary." Baltimore: Williams and Wilkins.

Taylor WA (1957a): Rothmund's syndrome—Thomson's syndrome. Arch Dermatol 75:236–244.

Taysi K, Say B, Firat T, Gürsu G (1971): Oculodentodigital dysplasia syndrome. Acta Paediatr Scand 60:235–238.

Thadani KI (1921): A toothless type of man. J Hered 12:87–88.

Thodén C-J, Ryöppy S, Kiutunen P (1977): Oculodentodigital dysplasia syndrome. Report of four cases. Acta Paediatr Scand 66:635–638.

Thurnam J (1848): Two cases in which the skin, hair and teeth were very imperfectly developed. Proc R M Chirurg Soc 31:71–82.

Tipton RE, Gorlin RJ (1983): Growth retardation, alopecia, pseudo-anodontia and optic atrophy—The GAPO syndrome. Report of a case and review of the literature. Am J Med Genet (in press).

Toriello HV, Lindstrom JA, Waterman DF, Baughman FA (1979): Re-evaluation of CHANDS. J Med Genet 16:316–317.

Toro-Sola MA, Kistenmacher ML, Punnett HH, DiGeorge AM (1975): Focal dermal hypoplasia syndrome in a male. Clin Genet 7:325–327.

Touraine A (1936): L'"anidrose avec hypotrichose et anodontie" (polydysplasie ectodermique héréditaire). Presse Méd 44:145–149.

Touraine A (1952): Les états héréditaires d'atrophies cutanées avec sénescence prématurée. Ann Dermatol 79:446–451.

Tridon P, Thiriet M (1966): "Malformations Associées de la Tête et des Extrémités." Paris: Masson.

Tuffli GA, Laxova R (1983): New, autosomal dominant form of ectodermal dysplasia. Am J Med Genet 14:381–384.

Tuomaala P, Haapanen E (1968): Three siblings with similar anomalies in the eyes, bones and skin. Acta Ophthalmol 46:365–372.

Upshaw BY, Montgomery H (1949): Hereditary anhidrotic ectodermal dysplasia. A clinical and pathologic study. Arch Dermatol Syphiligr 60:1170–1183.

Vassella F, Emrich HM, Kraus-Ruppert R, Aufdermaur F, Tönz O (1968): Congenital sensory neuropathy with anhidrosis. Arch Dis Child 43:124–130.

Verbov J (1970): Hypohidrotic (or anhidrotic) ectodermal dysplasia—an appraisal of diagnostic methods. Br J Dermatol 83:341–348.

Verbov J (1975): Anonychia with bizarre flexural pigmentation—an autosomal dominant dermatosis. Br J Dermatol 92:469–474.

Visveshwara N, Rudolph N, Dragutsky D (1974): Syndrome of accelerated skeletal maturation in infancy, peculiar facies, and multiple congenital anomalies. J Pediatr 84:553–556.

Vittori F, Carbonnel Y (1976): Fait clinique. Le syndrome oculo-dento-digital (dysplasie oculo-dento-digitale). Pédiatr 31:593–601.

Voigtländer V (1979): Pili torti with deafness (Bjørnstad syndrome). Report of a family. Dermatologica 159:50–54.

Volavsek W (1941): Zur Klinik der Nagelveränderungen und Palmarkeratosen bei Syringomyelie. Arch Dermatol 182:52–57.

Wahrman J, Berant M, Jacobs J, Aviad I, Ben-Hur N (1966): The oral-facial-digital syndrome: A male lethal condition in a boy with 47/XXY chromosomes. Pediatrics 37:812–821.

Wajntal A, Epps RR, Mendonça BB, Billerbeck AEC (1982): Nova síndrome de displasia ectodérmica: Nanismo, alopecia, anodontia e cutis laxa. Ciênc Cult 34 (suppl):705.

Walbaum R, Dehaene Ph, Schlemmer H (1971): Dysplasie ectodermique: Une forme autosomique récessive? Arch Fr Pédiatr 28:435–442.

Walbaum R, Fontaine G, Lienhardt J, Piquet JJ (1970): Surdité familiale avec ostéo-onycho-dysplasie. J Génét Hum 18:101–108.

Waldrigues A, Grohmann LC, Takahashi T, Reis HMP (1977): Ellis-van Creveld syndrome. An inbred kindred with five cases. Rev Bras Pesqui Méd Biol 10:193–198.

Wallace HJ (1958): Ectodermal defect with skeletal abnormalities. Proc R Soc Med Edinb 51:707–708.

Wannarachue K, Hall BD, Smith DW (1972): Ectodermal dysplasia and multiple defects (Rapp-Hodgkin type). J Pediatr 81:1217–1218.

Warkany J (1971): Syndromes. Am J Dis Child 121:365–370.

Weaver DD, Cohen MM Jr, Smith DW (1974a): The tricho-rhino-phalangeal syndrome. J Med Genet 11:312–314.

Weaver DD, Graham CB, Thomas IT, Smith DW (1974b): A new overgrowth syndrome with accelerated skeletal maturation, unusual facies, and camptodactyly. J Pediatr 84:547–552.

Weech AA (1929): Hereditary ectodermal dysplasia (congenital ectodermal defect). A report of two cases. Am J Dis Child 37:766–790.

Weiss H, Crosset AD Jr (1955): Chondroectodermal dysplasia. Report of a case and review of the literature. J Pediatr 46:268–275.

Weiswasser WH, Hall BD, Delavan GW, Smith DW (1973): The Coffin-Siris syndrome: Two new cases. Am J Dis Child 125:838–840.

Wesser DW, Vistnes LM (1969): Congenital ectodermal dysplasia, anhidrotic, with palatal paralysis and associated chromosome abnormality. Plast Reconstr Surg 14:396–398.

Whelan DT, Feldman W, Dost I (1975): The oro-facial-digital syndrome. Clin Genet 8:205–212.

Wiedemann HR, Grosse FR, Dibbern H (1978): "Características das Síndromes em Pediatria. Atlas de Diagnóstico Diferencial." Transl. by Dr. Horst Fürstenau. São Paulo: Editora Manole (A second, German edition of this book was published in 1982: Das characteristische Syndrom. Blickdiagnose von Syndromen. Stuttgart: F. K. Schattauer Verlag).

Wilbur RD (1973): Hypohidrotic ectodermal dysplasia. An unusual presentation. Med Ann Distr Columb 42:605–608.

Wilkey WD, Stevenson GH (1945): A family with inherited ectodermal dystrophy. Can Med Assoc J 53:226–230.

Wilkinson RD, Schopflocher P, Rozenfeld M (1977): Hidrotic ectodermal dysplasia with diffuse eccrine poromatosis. Arch Dermatol 113:472–476.

Williams M, Fraser FC (1967): Hydrotic ectodermal dysplasia—Clouston's family revisited. Can Med Assoc J 96:36–38.

Wilson FM II, Grayson M, Pieroni D (1973): Corneal changes in ectodermal dysplasia. Case report, histopathology, and differential diagnosis. Am J Ophthalmol 75:17–27.

Winter GB, Simpkiss MJ (1974): Hypertrichosis with hereditary gingival hyperplasia. Arch Dis Child 49:394–399.

Winter RM, Baraitser M, Laurence KM, Donnai D, Hall CM (1983): The Weissenbacher-Zweymüller, Stickler, and Marshall syndromes: Further evidence for their identity. Am J Med Genet 16:189–199.

Witkop CJ Jr (1965): Genetic disease of the oral cavity. In Tiecke RW (ed): "Oral Pathology." New York: McGraw-Hill.

Witkop CJ Jr (1971): Heterogeneity in gingival fibromatosis. Birth Defects VII(7):210–221.

Witkop CJ Jr (1979): Gingival fibromatosis and hypertrichosis. In Bergsma D (ed): "Birth Defects Compendium." New York: Alan R Liss, 2nd ed.

Witkop CJ Jr, Brearley LJ, Gentry WC (1975): Hypoplastic enamel, onycholysis, and hypohidrosis inherited as autosomal dominant trait. A review of ectodermal dysplasia syndromes. Oral Surg 39:71–86.

Witkop CJ Jr, Gorlin RJ (1961): Four hereditary mucosal syndromes. Arch Dermatol 84:762–771.

Witkop CJ Jr, Hudson CD (1979): Hypodontia and nail dysgenesis. In Bergsma D (ed): "Birth Defects Compendium." New York: Alan R Liss, 2nd ed.

Witkop CJ Jr, Sauk JJ Jr (1976): Heritable defects of enamel. In Stewart RW and Prescott GH (eds): "Oral Facial Genetics." Saint Louis: CV Mosby.

Wright CS, Guequierre JP (1947): Pachyonychia congenita—report of two cases, with studies on therapy. Arch Dermatol Syphiligr 55:819–827.

Zabel BU, Baumann W (1982): Langer-Giedion syndrome with interstitial 8q- deletion. Am J Med Genet 11:353–358.

Zamith VA, Campos JVM, Chizzotti TB (1974): Rothmund's syndrome. Serum and urine aminogram. Study of a family. Rev Bras Pesqui Méd Biol 7:23–27.

Zanier JM, Roubicek MM (1976): Hypohidrotic ectodermal dysplasia with autosomal dominant transmission. Fifth Int Congress Hum Genet, Mexico, Comunication 273.

Zellweger H, Smith JK, Grützner P (1974): The Marshall syndrome: Report of a new family. J Pediatr 84:868–871.

Zergollern L, Laktić N, Schmutzer L, Puretic S (1974): Focal dermal hypoplasia. Dermatologica 148:240–246.

Ziprokowski L, Ramon Y, Brish M (1963): Hyperkeratosis palmoplantaris with periodontosis (Papillon-Lefèvre). Arch Dermatol 88:207–209.

Addendum

When the manuscript of this book was already with the publisher, the following information came to our knowledge:

1. *Ectodermal dysplasia-ectrodactyly-macular distrophy (EEM syndrome)*. A new autosomal recessive malformation/dysplasia syndrome of the subgroup 1-2, described by Shozo Ohdo, Kiyotake Hirayama, and Tamotzu Terawaki, 1983, J Med Genet, 20:52–57.

2. *Corneal changes-hyperkeratosis-short stature-brachydactyly-premature birth*. A new autosomal dominant malformation/dysplasia syndrome of the subgroup 2-3, described by Judith K. Stern, Mark S. Lubinsky, Daniel S. Durrie, and John R. Luckasen, 1984, Am J Med Genet, 18:67–77.

3. *Hypotrichosis and dental defects*. A new pure ectodermal dysplasia of the subgroup 1-2 with unknown cause was referred to us by Dr. Gilbert B. Côté (pers. comm., 1984).

4. *Cleft lip/palate-oligodontia-syndactyly-hair alterations*. A new malformation/dysplasia syndrome of the subgroup 1-2 with unknown cause to be described by B. Martinez R., M.M. Mase, M. Pinheiro, and N. Freire-Maia (MS, 1984).

5. *Bartsocas-Papas syndrome*. A new autosomal recessive malformation/dysplasia syndrome of the subgroup 1-3, described by F. Papadia, F. Zimbalatti, and C. Gentile La Rosa, 1984, Am J Med Genet, 17:841–847.

6. *Migratory ichthyosiform exanthema-retinal coloboma-mental retardation-seizures*. A new malformation/dysplasia syndrome of the subgroup 1-2 with unknown cause, described by J. Zunich, and C.I. Kaye, 1984, Am J Med Genet, 17:707–710.

7. *Dry skin and extranumerary nipples*. A new pure autosomal dominant ectodermal dysplasia of the subgroup 1-4, described by E.A. Chautard-Freire-Maia, and N. Freire-Maia, 1984, Ciênc Cult, 36(suppl.):810.

8. *Curry-Hall syndrome*. A new autosomal dominant malformation/dysplasia syndrome of the subgroup 2-3, described by S.D. Shapiro, R.J. Jorgenson, and C.F. Salinas, 1984, Am J Med Genet, 17:579–583.

9. *Coffin-Siris syndrome.* The first case of consanguineous (first cousin) parents of a patient with this syndrome (subgroup 1–2–3) was reported in Brazil during the 36th meeting of the Brazilian Society for the Advancement of Science (July 1984). The patient is a six-year-old girl, the second in a sibship of three. Cf. A. Richieri Costa, M.S. Gonzales, R. Monteleone-Neto, 1984, Ciênc Cult, 36 (suppl):781. With this information, Coffin-Siris syndrome may be due to an AR mechanism of inheritance; however, heterogeneity is not excluded (AD?).

With the information contained in the above nine items, the subgroups of ectodermal dysplasias turn out to have the following distribution: 1–2–3–4 - 23; 1–2–3 - 30; 1–2–4 - 6; 1–3–4 - 9; 2–3–4 - 1; 1–2 - 20; 1–3 - 12; 1–4 - 3; 2–3 - 8; 2–4 - 3; 3–4 - 1. Total, 116; with heterogeneities, 125. Now, the numbers of the conditions are as follows: AD 38–49; AR 33–54; XL 5–15; unknown cause 19–60. Note that $38 + 33 + 5 + 49 = 125$. The value 60 is replaced by 49 because there are now 11 conditions that are counted twice (cf. Chapter 15).

Index

243